"I have savored *Cat People*. It's just wonderful. What a treat! It is totally entertaining—amusing, charming, touching, and insightful—alternately producing nods, listen-to-*this* calls to my husband, spontaneous guffaws, and tears. . . . You and Margaret have done a fabulous job retelling your cat tales. I feel as though I know your cats—can see each of them in his/her favorite spot. More than telling the story, you and Margaret have done a fabulous job caring for and *loving* all these pets. I'm in awe of you for that."

—Barbara Delinsky, author of
Looking for Peyton Place

"*Cat People* proves once again that you recognize it's important to be humble before one's superiors, in this case your cats. Reading *Cat People* proves how very intelligent you are. Most important, you will impress your cat."

—Rita Mae Brown, author of
The Hunt Ball and *Cat's Eye Witness*

Chris Ramirez/WpN

About the Authors

MICHAEL KORDA is the author of many books, including *Horse People, Country Matters,* and *Ulysses S. Grant.* MARGARET KORDA was born in England and now lives with her husband in Dutchess County, New York. They are also coauthors of *Horse Housekeeping: How to Keep a Horse at Home.*

CAT PEOPLE

Margaret and Michael Korda

HARPER ● PERENNIAL

NEW YORK ● LONDON ● TORONTO ● SYDNEY

HARPER ● PERENNIAL

A hardcover edition of this book was published in 2005 by
HarperCollins Publishers.

HarperCollins books may be purchased for educational, business,
or sales promotional use. For information please write:
Special Markets Department, HarperCollins Publishers,
10 East 53rd Street, New York, NY 10022.

FIRST HARPER PERENNIAL EDITION PUBLISHED 2006.

Designed by Amy Hill

The Library of Congress has catalogued the hardcover edition as follows:
Korda, Margaret.
Cat people / Margaret and Michael Korda.—1st ed.
p. cm.
ISBN-10: 0-06-075663-2
ISBN-13: 978-0-06-075663-5
1. Cats—Anecdotes. 2. Cat owners—Anecdotes. I. Korda, Michael. II. Title.
SF445.5.K67 2005
636.8—dc22   2005046032

ISBN-10: 0-06-075664-0 (pbk.)
ISBN-13: 978-0-06-075664-2 (pbk.)

13   ❖ / RRD   10   9   8   7   6   5   4   3

For Tamzin

And in loving memory of Kit

When I observed he was a fine cat, [Johnson said], "why yes, Sir, but I have had cats whom I liked better than this"; and then, as if perceiving Hodge to be out of countenance, adding, "but he is a very fine cat, a very fine cat indeed."

—Boswell's *Life of Johnson*

We who choose to surround ourselves with lives even more temporary than our own live within a fragile circle, easily and often breached. Unable to accept its awful gaps, we still would live no other way.

—Irving Townshend, *Separate Lives*

CONTENTS

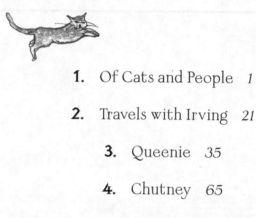

CONTENTS

1. Of Cats and People

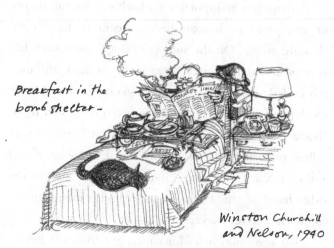

Breakfast in the bomb shelter —

Winston Churchill and Nelson, 1940

Cats and people have lived together for a very long time, ever since humankind turned from hunting and gathering, for the former of which pursuits dogs were useful, to settled agriculture, for which protecting the harvested grain and seed from rodents began to matter more.

Large members of the cat family were a predatory menace to early humans (and of course still are in parts of Africa, Siberia, and India), but the smaller members of the cat family no doubt made themselves noticed around the campfires of our remote ancestors by their proficiency at killing mice and rats.

Perhaps just as important for both species, cats do not at any point in human history seem to have been thought edible. On the subject of dogs, tastes vary, but American Indians prized them highly as food, and usually boiled a puppy as a treat for esteemed visitors; Polar explorers in an emergency have eaten their sled dogs, though no doubt with regret; and dogs still play a part (albeit one that most Americans and Europeans would rather ignore) in Chinese haute cuisine. Cats, on the other hand, are not normally on anybody's menu, and so far as one can tell, never have been.

That was good news for cats, right from the beginning. Not, by nature, particularly trusting creatures—in the wild, they are amiable, curious, but cautious—at least in their relationship with humans they were spared the pot, except in the extreme circumstances of great sieges, in which besieged city-dwellers were reduced to eating dogs, rats, and their own boots, before turning to cats as a last resort.

Since cats made themselves useful by killing rodents, exercised, groomed, and fed themselves, and were, as animals go, exceptionally clean and tidy in their habits, even fastidious, it is hardly surprising that humans accepted them quickly into their domestic environment, surely noting, from the beginning, that cats were not on the whole looking for a leader or master, and indeed had an independent and fairly haughty view of themselves, and their place in the world.

Dogs are animals of the pack, eager to follow a leader, and even willing to accept a human being in that position, provided he or she feeds them. Cats are independent—leadership is not high on their list of demands—and it must have been quickly apparent that no cat was likely to look up at a human being with an adoring, trusting, or soulful expression in its eyes, or do anything on command.

Dogs can also be taught a great deal, and most of them enjoy learning—certainly they enjoy being praised—but cats have no interest in learning at all, what they do is done by instinct, and they do not seem to feel they have anything to learn from people, nor does their day generally depend on whether they have been praised or not. Cats don't do tricks to please their owner—if they do anything resembling a trick, it's to please and amuse themselves.

It is hardly surprising that the Egyptians, with their economy dependent on intensive agriculture and storage of grain, came early on to treat cats as godlike figures. Aloof, beautiful, mysterious, able to see in the dark, fierce but apparently passionless killers, who toyed with the creatures they intended to kill as the gods toyed with humans' lives, cats played a huge role in Egyptian iconography and art, far beyond their importance as rodent killers in the grain storage bins of the kingdom. The cat-like eyes, slender figures, and long necks in the portraits of the Egyptian queens make it clear enough that cats were worshipped and imitated for their beauty as much as for their usefulness. The highest form of human beauty in ancient Egypt was to resemble a cat, and cats themselves sometimes wore collars of gold and precious stones.

Of course the cat's nature does not inspire universal love. Those who crave obedience from an animal are unlikely to appreciate cats as pets—they do not heel, or roll over and play dead, or hold out a paw on command—and the world since ancient Egypt has therefore been divided between cat lovers and those who can't abide cats. Not that cats mind much. They have their own agenda, and like to stick to it.

Perhaps this spirit of independence, and the fact that they remain, even when domesticated, essentially wild, is what gives them their remarkable powers of survival—the proverbial nine lives. When it comes to independence, survival skills, loyalty to your own kind, and unconditional love, we humans have much to learn from cats.

We live in rural Dutchess County, where winters are

long and hard, and there is never any shortage of stray or abandoned cats in the woods and fields around our house. Many of them look scrawny, underfed, and resentful—one guesses that most of them have been thrown out of a comfortable life in somebody's house or trailer into the wild for one reason or another—but a surprising number of them survive, as we shall see, living in drains, or old barns, and foraging for their food.

Dogs, of course, can survive in the wild too, except perhaps for the smaller, more decorative species, but they usually need to form or join a pack to do so, whereas cats seem to slip effortlessly back to wildness when they have to, leading solitary lives in harsh conditions. Most of our cats, in fact, have emerged from the woods to the front door or the porches, become accustomed to finding a bowl of food put out for them, then made the transition from strays, to "outside" cats, who hang around the property, then to becoming "barn cats," who live in the barn and do a little light mousing and ratting to earn their keep, and finally get promoted to "house cat" status, settling down on the chairs and sofas as if they had never roamed wild through the winter or hunted for their dinner.

Some cats, however, seem to have a problem deciding whether they want to come in from the cold at all. They will come close to the house at night, if food is put outside for them, but resist any attempt to bring them indoors, even in below-zero weather.

In general, we have never sought out cats—they have sought *us* out, after a good deal of observation, as if they were "casing the joint." Some of our cats have been daily visitors—or been seen daily in the fields while we were riding—for months on end before finally deciding to present themselves at the house for adoption. Others accept a meal, but stay aloof.

At this moment, for instance, as we write, with snow, ice, and frozen rain on the ground (and more coming), there is one male cat out there somewhere on our property, whom we have named Agent Orange (because of his color), who turns up regularly for his meal at Stonegate Farm, but doesn't seem to want to be caught. For a while he had competition, in the form of a female cat named Tizz Whizz, but she eventually accepted the tack room in the barn as her residence. When they were both living outdoors, they seemed to have divided up our property into independent sectors—the front door porch was his spot for his evening meal, while the aisle in the barn was hers. They came and went on different schedules, so as not to run into each other, presumably. You would suppose that teaming up to survive might make sense, but that's because you're not a cat.

Of course it's possible to *buy* cats, or adopt them from the ASPCA or similar organizations, rather than picking up strays, but living as we do on a farm that's never been necessary for us since the day we moved up here with our two city cats. There are always plenty of cats out there, some desperately looking for a home, others, perhaps based on previous experience with human behavior, not at all sure that they want to exchange freedom for domesticity. In any event, however plump, cosseted, and contented a "house cat" may look snoozing on your best chair in front of the fireplace, it is still, at heart, a wild animal, and with whatever regrets for past comfort, able to look out for itself like one if need be.

This may be another reason that some people dislike cats. Even sitting quietly on your lap, claws sheathed, a part of them is still wild. They may have come indoors, and agreed to accept your caresses, but there's always the sense that one eye is ever so slightly open and wary, just in case the cat changes its mind. If there's one thing you can say about cats, they like to keep their options open—not a bad thing to learn from them.

Still the main thing separating dog lovers from cat lovers is the formers' ideal of a pet—one which pines when you're away, rushes to greet you when you come home, and slavishly admires everything you do—in contrast to the cat lovers' fondness for a pet that ignores your comings and goings, shares your home pretty much on its own terms, and displays affection when it feels like doing so, and not a moment sooner. Not everybody can bridge that gap.

Also, cats are great energy conservers. Eighteen hours of sleep a day (interrupted by a couple of meals) is nothing to a cat. A dog may share your morning run with great enthusiasm, but a cat is only likely to stretch a little, open one eye a crack as you leave, then roll over and go back to sleep. Cats get their exercise in short, sharp bursts, usually when you're not looking, or are asleep yourself. We have a cat that likes to watch when Michael does his yoga, perched comfortably on the back of a sofa unraveling the upholstery with her claws, but she has no desire to join in. Like most cats, she is as fit as she wants to be.

A good lesson for us all?

We came to cats by different ways. Margaret grew up on a farm in the English countryside, surrounded by cats, with a cat-loving mother, and a father who, like many farmers, was not a sentimentalist about small animals.

She recalls an example of this in the early summer of 1959. She and her mother, Kit, were sitting on the lawn, her mother in the remainder of the shade, watching the trout in the mill pond come up for insects, making soft, splashing sounds. Her father had gone back out after tea to look at something on the farm, driving off in haste and leaving a cloud of dust behind him, which irritated her mother a good deal.

"Let's catch some trout for your father's supper, and surprise him," Kit said.

"We haven't got anything to catch them with."

"Well, we'll go and have a look, won't we, lovey?"

And of course they did.

As Margaret remembers it:

"She found a safety pin. Twisted it out of shape, tied a piece of string to it, and got some bacon out of the fridge. Not happy with one, she fashioned two. We leaned over the heavy metal rail and dangled our lines. Within minutes we had two trout hooked and flopping on the lawn. 'Oh God, we don't know how to kill them,' I said.

"'No, but I think we can get the hooks out. You get some dish towels from the kitchen, and when they have flipped themselves into the shade, cover them, and we won't have to watch them dying.'

"But they never appeared for my father's supper.

"'What happened to them?' I asked her in the kitchen later. 'You cleaned them.'

"'I cooked them up for the cats, they loved them,' Kit said.

"'But . . .'

"'What the eye doesn't see, the heart doesn't grieve for.'

"And that was exactly the same answer I got when I found her one day filling a large roasting pan with dirt from the garden, to serve as a litter box. 'It's for the new kittens, they'll become house-trained in no time.'

"'Better get a new one before we have our next roast dinner,' I suggested.

"'No, no, lovey. A good wash out with boiling water, it'll be as good as new. And don't worry about your father, he'll never know. What the eye doesn't see . . .'

"But my father's eye missed nothing, saw everything. A few weeks later when my mother went to lose a few pounds at a fashionable spa, he had the gamekeeper come by and shoot them all."

Michael, on the other hand, grew up in cities, far from this kind of rural drama, with a father who always owned a dog, but usually liked to have a cat around as well. The dog went to the film studio with him every day in the car—Vincent was an Academy Award–winning art director—while the cat stayed put in the house and sat on his lap in the evening when he came home to read his paper and drink a few glasses of wine in front of the fire.

An old photograph of Vincent in his earlier days as a painter in Montparnasse, shows him sitting next to a friend and fellow painter; each of them is holding a cat, and Vincent is looking rather enviously at his friend's, as if he had picked the wrong one. Like most painters of his time he sometimes included a cat in his portraits—cats have the great advantage for a painter that they are usually willing to sit motionless for long periods of time, which is exactly what you want in a model.

Another nice thing about cats is that they seem immune, or even inimical to the "great man" syndrome. When Richard Nixon's advisers were trying to improve the president's image, they recommended getting him a dog, and even chose the breed after much research, a red setter named King Timahoe. (The strategy backfired, as King Timahoe hated Nixon and growled whenever he saw him, and Nixon's aides were obliged to keep dog biscuits in one of the drawers of his Oval Office desk so the president could tempt the reluctant dog within petting distance for the occasional photo op.) As you may imagine, nobody advised Nixon to get a cat.

It's probably just as well. Generally, those who take

themselves seriously tend not to be cat lovers, since cats don't take anyone seriously. It is hard to imagine Hitler with a cat, for example, instead of Blondi, his faithful German shepherd, whereas Winston Churchill adored cats and on becoming First Lord of the Admiralty in August 1939, adopted the Admiralty cat (named, unsurprisingly, Nelson) as his own when he and Mrs. Churchill moved into the First Lord's flat. Nelson, a bulky and dignified black cat, used to sleep on Churchill's bed, sprawled at his feet, and Churchill would pet him thoughtfully while he dictated messages or held bedside conferences, and sometimes ask Nelson what he thought of the issues being discussed, to the great annoyance of the admirals.

When Churchill moved to 10 Downing Street, after becoming Prime Minister in May 1940, Nelson left the Admiralty to go with him, and distinguished visitors to the Prime Minister's bedside were always formally introduced to him, and privileged to watch Churchill feed him scraps of bacon off his breakfast tray. Every once in a while, Churchill would glance at Nelson affectionately and say, "Cat, darling."

Poor Nelson was afraid of the sound of London's antiaircraft guns, and during air raids the prime minister would remind him of the hero whose name he bore.

"Try and remember," he told Nelson sternly one evening, to the amusement of Anthony Eden, "what those boys in the R.A.F. are doing."

When it was necessary for Churchill to go down to the air-raid shelter, he always took Nelson with him, having made sure that appropriate arrangements had been made for the cat.

Churchill's private secretary John Colville captured the Prime Minister's affection for the ex-admiralty cat—as well as Churchill's peremptory style and phenomenal nervous energy—with a widely circulated spoof of a typical Churchill memo, written during the darkest days of the Blitz:

> 31 October 1940
> ACTION THIS DAY
> Pray let six new offices be fitted for my use, in Selfridge's, Lambeth Palace, Stanmore, Tooting Bec, the Palladium, and Mile End Road. I will inform you at 6 each evening at which office I shall dine, work and sleep. Accommodation will be required for Mrs. Churchill, two shorthand writers, three secretaries, and Nelson. There should be shelter for all, and a place for me to watch air raids from the roof.

It would be easy to take the view that the good guys like cats, and that is to some degree borne out by other examples. A visitor to the White House in the darkest days of the Civil War was surprised to find Lincoln stretched out in front of the fire with his shoes off, while a number of kittens ran up and down his long legs, tumbled in and out of his pockets, and perched on his shoulders, but then it was part of Lincoln's charm that he did not take himself altogether seriously, and retained a certain playfulness of spirit, to which the kittens must certainly have appealed. There are no descriptions, it is almost needless to say, of Caesar, Napoleon, Hitler, or, for that matter, Jefferson Davis playing with kittens.

Ships, of course, have always had a ship's cat since man first went to sea; the cats led privileged lives aboard in exchange for killing rats. Unlike dogs, cats, it seems, do not suffer from seasickness, and at least in the Royal Navy and the British Merchant Marine they were treated with the respect due to a full member of the crew, and considered lucky mascots as well. In wartime captains might go down with their ships—Admiral Tom Phillips did when he lost two British battleships, H.M.S. *Repulse* and H.M.S. *Prince of Wales,* to Japanese bombers off the coast of Malaya shortly after Pearl Harbor, but the two ships' cats survived—and even today it is usual to find a place for the ship's cat in one of the lifeboats, and indeed considered bad luck (and poor seamanship) not to do so and leave the ship's cat to go down with the captain.

Although there is no official place for cats in the British armed forces, except for the Royal Navy, any army barracks or R.A.F. station is sure to contain a surprising number of cats. No kitchen or mess is complete without a cat-in-residence, and soldiers and airmen adopt cats, even when strictly forbidden to do so. Soldiers going into the trenches in the First World War often hid stray kittens in their tunics, fed them from their rations, and found a certain peace of mind in comforting the animals during bombardments. (Cats were, as you might guess, highly esteemed in the trenches, given the number of rats in them.) They maintained their appeal during the Second World War—more than one soldier landed on the Normandy beaches hiding a cat, at least one British paratrooper dropped into Arnhem with a cat zipped into his jumpsuit, some tank crews are reported to have kept a pet cat in the turret of their tank, and one intrepid cat accompanied his owner, a Royal Air Force pilot, on bombing raids over Germany, snoozing on a parachute pack despite the flak.

That's the way it is—once you let a cat into your life it

is likely to share the best and the worst of it with you. It's not so much a question of loyalty, as of habit. Besides, cats are adventurous, and born survivors, just as capable of bringing up a family in a subway tunnel (or a tank turret) as in the finest of homes, if need be.

Generally speaking, however, cats prefer to observe human activity from the prone position, with a skeptical eye. Either you find that comforting, or you don't. It's all the same to the cat.

As for a spirit of adventure, on that score it's hard to beat a cat. Michael's initial experience of living with a cat as an adult was during his first marriage, when his then-wife, Casey, suddenly announced out of the blue that she wanted a cat—not only did she want one, but, by God, she was going to *have* one (she was a Bennington girl), and what was more she knew just what kind she wanted—a Burmese. Since they then lived in a small apartment on East 56th Street, with their son Christopher, he thought that quarters would be a little cramped with an additional resident, however small and furry, but consoled himself with the thought that this, like many of Casey's enthusiasms, would pass.

Needless to say, it did not. A city girl, she went to Fabulous Felines, then a well-known pet shop on Lexington Avenue, wrote out a check for a Burmese kitten, and brought it home, together with its water bowl, feeding bowl, scratching post, and litter tray. Like most cats, it took one quick stroll around the premises—not difficult since they were so small—shrugged, and settled down on the couch. As kittens go, it was young enough to miss its mother, but cats and kittens don't grieve or mope much, and tend to accept the hand that fate has dealt them stoically. Napoleon—as the kitten was quickly named—had his namesake's piercing eyes and masterful disposition, and despite his initial objections, Michael soon became accustomed to him, and he to Michael.

They lived on the fourth floor of a big apartment building, next door to the Midtown Tennis Club, and during Christopher's infancy had taken the trouble to put screens on most of the windows, so, once he started walking, he couldn't fall out. Not all the windows had been screened—you had to remember which ones were safe to open and which were not—and this became doubly important once Napoleon joined the family, as cats are on the whole more likely to jump up on windowsills than children, and if a window is open are apt to keep going.

One evening, Casey was cooking something in the kitchen in the toaster oven that caught fire—haute cuisine had apparently not been one of the things they taught at Bennington—and the apartment quickly filled with smoke. Anxious not to choke to death, or ruin all the furniture with greasy smoke, Michael opened a window, then rushed into the kitchen to help put out the fire. When the smoke began to clear, Napoleon was nowhere to be found, and after a quick search, his eye caught the window, and Michael realized to his horror that he had opened one of the windows that *wasn't* protected by a screen!

He closed it quickly, but it was all too clear that the tragedy had already occurred, and a closer search of the apartment made it clear that Napoleon had not gone into hiding, as he hoped, but had gone out the window into the night, four floors down onto 56th Street.

Glumly—he had plenty to be glum about, on top of a natural feeling of guilt and sadness—he went downstairs with a flashlight to look for the mangled corpse. He explained what had happened to the doorman, who sighed sadly, and said he had seen nothing and heard nothing. His mood was not improved by the weather—it was pouring rain (raining "cats and dogs" as they say, he reminded himself lugubriously), and he sloshed up and down 56th Street looking for the tiny body, perhaps further mangled and flattened by a passing taxi. There was no sign of it, however, and he knew better than to come home without it. Thinking that the cat might have taken a slightly different trajectory, he talked his way past the night watchman onto the courts of the tennis club, but there was nobody there either. He stood in the dark for a few minutes, holding his flashlight and trying to think of something sensible to say when he went back upstairs, but nothing useful came to mind. He swung his flashlight around, taking one last look.

Then, all of a sudden, he noticed two yellow eyes staring down at him from the roof of the tennis club. There was no proof they were Napoleon's—New York is full of stray cats—but slipping the night watchman five dollars, he persuaded him to let him climb up the fire escape, where, soaked, but otherwise unharmed, he found Napoleon. He picked him up and climbed back down again. "That a mighty lucky cat," the watchman said, shaking his head.

Michael thought so too. As he looked up, he could see what had probably happened. Napoleon had stepped out of the window, dropped four floors down, landed on top of the building's cloth canopy, thus breaking his fall, bounced on the heavy canvas and took a flying leap to the top of a dividing brick wall, then jumped down about twelve feet onto the tennis club.

Four stories is a long fall—at least forty feet, more than enough to kill a human being—but Napoleon wasn't even sore or lame, just wet and mildly irritable.

Michael brought him home to his grieving wife and son, who were so astonished to get him back alive that Michael was instantly forgiven for having opened the wrong window, and came away from the experience with a deep admiration for a cat's survival instincts, and a phobia about open windows that he never lost.

When Michael's marriage broke up—perhaps the incident had gone deeper than he assumed at the time—Margaret, the woman he had fallen in love with, turned out to be a cat lover too. Indeed, her cat, a large orange male named Irving, was the witness to their love affair, and it was clear, first of all, that he did not approve of Michael—he didn't like her husband any better—and secondly, that he came along with Margaret, wherever she went.

He always had.

And he always would.

2. Travels with Irving

"Live and let live"—
Irving confronts a mouse

I saw Irving in a pet store window on Columbus Avenue," remembers Margaret. "There were several other kittens but he sat to one side on his own. He was orange and looked vulnerable. I went in and bought him and carried him home inside my coat. We made lots of small excursions together in the beginning around New York. Down to the drugstore, the supermarket, my friend Mayo's apartment, to visit the vet, to Bloomingdale's, and to the occasional movie.

"As time went by and we became inseparable—not a healthy thing, my then husband said when he was around, which was not often—we made longer and more adventuresome trips. Sometimes with him too, a trio. I remember staying in a suite at the Beverly Wilshire Hotel in Los Angeles. I had called in advance to inquire about their policy toward having pets in the rooms. The person to whom I spoke said they allowed small dogs and couldn't see that a cat would be much different. They obviously knew little about cats! Irving had a great time with the floor-to-ceiling drapes, especially the lining, which seemed irresistible to him. When we checked out we were asked not to bring the cat back again. Burt, my husband, seemed a little unnerved—he was, in his world, somewhat of a celebrity—and unaccustomed to negativity at check-out time, quite the opposite in fact, eager hotel staff wanting to know the date of a possible next visit. He approached Irving and me, in a fog of cigar smoke, cameras and light meters swinging from his neck. 'What the fuck did he do? I'm telling you, kiddo, it'll take a miracle to get back in here next time.'

"We came back to New York on American Airlines, and in those faraway days, they had an area in the rear of the plane called the Piano Bar. It was cozy and Irving came out of his travel box and spent the flight on my lap.

"In the early seventies we took a house every winter in Cuernavaca for a month. I was not going without Irving and announced this at one of our black-tie dinners for twelve, among which were several couples who were planning to spend some time with us. The chatter stopped, eyes swung around to stare at me, Burt inhaled on his cigar, and when he could speak, he said, 'Never going to happen, kiddo, the airlines, the Mexican authorities, getting him back into this country, you name it, no way.' Some guests, anticipating a dinner table squabble, added their thoughts. What about cat food, litter, how would he travel? My friend Mo, who was going to drive down from Washington with her husband, said quietly, 'Oh, we're taking the food and litter, whatever he needs.' 'No problems with traveling,' I said. 'I already called Air France, they have the best flight, stops in New York en route from Paris to Mexico City, and they said they did not care what I brought along, provided I got it onto the plane, I could bring a cow if I wanted. And I thought I could find some sort of collapsible cardboard box to use as a litter tray during the flight, and Burt could put a plastic bag of kitty litter in one of his carry-on camera bags.'

"There were some shifty looks around the table and a nervous laugh from one guest. 'I'm not carrying any goddamn cat shit stuff in with my cameras, kiddo, are you crazy?' Burt said.

"'We'll see. Let's have coffee, shall we?'

"Irving went to Mexico several times.

"He also went to the Okefenokee Swamp for a long weekend. We traveled by overnight train from New York, via Washington where John Chancellor, who was then the anchor for NBC's *Nightly News,* and his wife, Barbara, joined us, arriving at Mekong, Georgia, the next morning. I didn't take Irving out on the boat in the swamp, a bit too risky, but he loved the train ride, sitting on top of the seat above my head, watching the country roll by. We went to Lexington, Kentucky, and up to Martha's Vineyard. But my life changed and I traveled less and so did Irving.

"In the end, once Burt and I were divorced, and Michael and I bought a farm in the country, Irving's trips were reduced to driving back and forth to Dutchess County, where he finally settled. For all his travels, he was an indoor cat and unused to the outside. I often wonder what he thought, going from pet shop window to 'trains, planes, and automobiles,' and finally from an apartment on Central Park West to quiet retirement in a country house."

Irving had his faults, but he was the most faithful of cats, and totally devoted to Margaret, to the exclusion of all other interests. By the time Michael entered his life he was a cat of firmly fixed habits, and used to life as an apartment dweller. When we bought our house in the country, as a weekend retreat at first, Irving disliked the drive there and back almost as much as he had disliked Burt's cigar smoke. He did not suffer in silence—he yowled, drooled, moaned, and threw up all the way from Central Park West and 65th Street to our farm on Friday nights, and did the same on the way back on Sunday nights. Very often, he threw up before we had even left the garage or the driveway. It sometimes seemed as if he spent the entire week living in dread of the two-hour drive to Dutchess County.

Saying, "Oh, well, he'll get used to it" (as we did several times every weekend), about any cat, by the way, is generally a mistake. Cats seldom get used to things they dislike. Like the Bourbons, on their return to France after the long years of exile during the Revolution and Napoleon's empire, of whom Talleyrand said, "They forgot nothing and they learned nothing," cats have a good memory for anything that has been done to them that they resent or dislike, and very little capacity for forgiving and forgetting. Nor are they easily bribed. With cats, first impressions count, and the cat's initial meeting with a person is likely to imprint itself on its mind permanently.

In Irving's case, age, or perhaps disappointment that when Margaret left Burt, Michael would eventually arrive to replace him—Irving would certainly have preferred to have had Margaret to himself, and never made any secret of the fact—had soured him on travel. After all, at one time in his life he had been a well-traveled cat and not even the longest of trips had dismayed him, so it's hard to see why he should have taken such a dislike to a piddling little two-hour commute to the country. Or perhaps it was just that he was used to traveling first class on airplanes, and being fussed over by flight attendants, rather than being chucked onto the backseat of a car, together with his box and a litter tray, and regarded it as a distinct comedown in the world, having traveled to the Okeefenokee Swamp by train in a private sleeping compartment and stayed in a suite at the Beverly Wilshire Hotel. In those easygoing days, airlines accommodated easily to cats, at any rate in first class, and Irving flew the skies with his own litter tray, dinner dish, and water bowl, back when airplanes had piano bars, and Braniff stews wore miniskirts and patent leather boots designed by Courrèges. In short, Irving was used to the best, and no doubt brooded wistfully over the days when he had roamed the world with Margaret in style.

And why not? After all, cats have standards too. As easily as they revert to the wild, they get used to a certain pattern of comfort. The wrong kind of cat food, a change in the way the furniture is arranged that eliminates a favorite place to nap, a break in the household routine, is more than enough to send them into a snit, or give them a bad case of the sulks. What's more, they are skilled at making their "owner" (Does anybody really "own" a cat?) feel guilty. Cats may not be able to smile, but they certainly know how to look aggrieved, and can resist or ignore any attempt at reconciliation for a very long time indeed. Nor are they easily bribed. Stroking them, scratching them behind the ears, offering them a particularly succulent treat, will rarely do the trick. Only a prolonged campaign of apologies will bring forgiveness, which won't come until the cat is good and ready to offer it, and not a moment sooner.

Cats rarely attack people except in what they see, no doubt, as self-defense (though there *are* exceptions, as you will see when you meet a cat of ours named Mrs. Bumble, named after the object of the parish beadle's affections in *Oliver Twist*), but even the nicest of cats will signal displeasure by a quick swipe with the claws, when all other measures have failed or been ignored, and we keep a bottle of Mercurochrome or iodine on hand upstairs in the bathroom and downstairs in the kitchen, and of course a box of Band-Aids, for just such moments. Ours mostly express dissatisfaction by means of a series of graduated acts of low-level vandalism, designed to attract attention. First they will knock over small picture frames, then, if that fails, they scratch the screens—a noise which resembles that of a fingernail on a blackboard after a few minutes. Then they move on to heavier stuff—sharpening their claws on expensively upholstered furniture or the better rugs and carpets, pushing over bowls of flowers or their own water bowl, making a diligent attempt to scrape the wallpaper off the walls, etc. By that time, no matter how determined we have been to sleep in a bit on a Sunday morning, we are both up and ready to go downstairs and open cans of cat food. And that's forgetting such guaranteed waker-uppers as prolonged yowling or throwing up on the faux-oriental carpets.

Smaller and younger cats can chase each other at high speed around the bedroom floor, leaping up and down off the bed, and landing on furniture with a terrific thump that dislocates every item on it and sends glasses, pens, and books flying, but the older cats don't have to bother with physical exertion—a good, long session of strumming tunelessly at the screens with a claw will do just as well, they have learned, if not better, and hardly requires any effort at all.

Well, living with cats is a compromise, after all, and most of the compromising inevitably has to be made on the human side, since cats don't compromise easily, and tend to regard your turf as theirs. Cats have a genuine sense of entitlement—even when they are on the outside looking in, as strays, most of them do not stoop to looking wistful or appealing; rather than grovel like dogs, their approach is usually demanding. A photograph of one of Margaret's all-time favorite cats, Jake, looking in through the kitchen window on a winter's day makes that clear. The expression on his face is haughty, mildly impatient, and slightly censorious, as if he were saying, "You can *see* I'm out here in the cold waiting to be let in, how long do I have to wait here?" One look at him is enough to tell you that there will be no fawning gratitude from him when the door is finally opened. If he could speak, he would no doubt say, "About time, too!," and stomp off grumpily to sharpen his claws on the dining-room carpet.

Cat people, almost by definition, have learned to compromise with their cats, which is sensible, since the cats are unlikely to do much in the way of compromising back. But this is the feline way, and perhaps part of the cat's appeal to human beings, who are generally used to getting their way with animals. You can search through history, newspaper files, and literature in vain for the cat equivalent of the traditional heroic dog story—the feline equivalent of Lassie, or Rin Tin Tin, or Balto (the brave sled dog that brought the serum to Nome, Alaska, and who is immortalized in a life-size statue in New York's Central Park, as well as by a popular brand of French cigarette), or the Seeing Eye dog, simply does not exist. There are stories aplenty of cats who find their way home from hundreds of miles away (usually because they have been dumped or forgotten when the family moved), but none of a cat bravely sacrificing its life to save its owner's, or plunging into the flames to rescue a child, or being used by the police to detect narcotics or explosives.

The standard newspaper story about cats, quite the contrary, is when the fire department has to be called out to rescue a cat that has climbed so high up a tree that it can't get down, and even then, it usually manages to scratch its rescuer once it's been reached by ladder—in short, the typical cat story is when a lot of valuable equipment and the time of a whole lot of firefighters, police officers, and animal protection officers are spent to rescue a cat from someplace where it shouldn't have gone in the first place, and get no thanks for their trouble from the victim. Humans go to large-scale heroics on behalf of cats, but cats generally don't go in for heroics on behalf of humans, and that's that—it is hard to imagine that even a whole room full of cats would attack a burglar. Judging from ours, they would open one eye, then go back to sleep—certainly they would not guard the family silver or the jewelry box or the gun cabinet with teeth bared. On the other hand, cats *will* go back into burning buildings to rescue kittens—they are quite capable of heroism on behalf of other cats.

Even geese are better at guarding things than cats—as the Romans knew, when they surrounded the capitol with geese, which would let off an ungodly row of cackling and honking at the approach of a stranger at night. Of course geese have their downside, as anybody who has ever stepped in goose shit can testify, and neither their bathroom habits nor their quarrelsome tempers make them easy to keep as pets, but if you want animals that will sound the alarm loud and clear and attack an intruder fearlessly, geese might be worth considering—

a full-grown goose coming for you in a rush, wings flapping and beak snapping aggressively, is something before which even the staunchest burglar might retreat.

But cats don't do that. Admittedly, some of their bigger relatives—lions, tigers, leopards—are as fearsome as an animal gets, but your average domestic cat is definitely not a substitute for an attack dog, though there are cases on record of cats waking up their owners when the house is on fire or a burglar has broken in.

Cats are by no means simply useless or decorative animals. If you happen to have a mouse around, most cats will do their best to kill it for you (though you will not necessarily appreciate the mess they make, since cats are neither neat nor tidy killers), but even mousing, when performed indoors, is pretty much dependent on the cat's appetite (outdoors they may do it for *le sport*). Given two squares a day and a frequently filled bowl of dry food for between-meal snacks, a cat's desire to kill mice diminishes rapidly to zero.

Our own have actually been seen to nap while a mouse darts boldly across the room in front of their noses. They may open an eye to give it a curious glance, but if they're not hungry, they're apparently willing to live and let live, and the mouse seems to know it. When we first moved to the country the equation seemed to us simple: cats = no mice. But that has not proved to be the case. We have, over the years, invested countless dollars in electronic signals that are either supposed to drive mice mad or away, mousetraps from the most sophisticated to the simple, old-fashioned ones (baited, according to local custom, with peanut butter or Kraft cheddar singles), and various kinds of poison that are supposed to be harmless to domestic animals, not to speak of lining the drawers with sheets of tin. But still, in the long winter nights, we can hear the mice scampering around in the walls and overhead in the attic, while the cats sleep soundly, indifferent to the patter of innumerable tiny feet. "Nothing to do with me," seems to be their motto, when it comes to mice. Or birds in the house. Or bats.

Our cats, it will be surmised, *share* our lives, pretty much on their own terms, take it or leave it. We may see ourselves, from time to time, as having played a benevolent role by giving them a decent home and regular meals, but the cats don't seem to see it that way at all. On the contrary, they see *us* as the privileged ones, whom they have honored with their company. And perhaps they're right—humans can learn a trace of humility from the way cats see them. No bad thing!

Every once in a while one reads of another chapter in the long debate of cats vs. dogs. A dog lover had written in to a pet magazine determined to prove that dogs are smarter than cats, one of her illustrations being that whereas dogs quickly learn to recognize their own name and respond to it, cats don't. But is this really so? Nonscientific observation seems to indicate that cats do indeed recognize their own name, but unlike dogs, feel it's beneath their dignity to respond to it. After all, our name for a cat may not be *its* name for itself, to begin with—just because you've decided to call her "Tulip" or "Kit Kat" doesn't mean the cat has accepted that name—then too cats *do* respond to the name they've been given by humans, but only if it's to the cat's advantage. We can stand at the front door at night shouting "Hooligan!" over and over again until we're hoarse, without seeing a trace of Hooligan (who being jet black simply vanishes when it's dark), but if you add to her name the magic words "din-din" (for dinner, needless to explain), Hooligan will emerge out of the night into the light of the front porch in a few seconds. She recognizes her name all right, but unlike a dog she's not about to coming running at the sound of it—unless, of course, it's accompanied by the promise of dinner.

From which one might conclude that, in fact, cats are smarter than dogs, and also don't make a fetish of obedience.

Ours sometimes seem very smart indeed.

3. Queenie

Queenie

Our dear friend Larry Ashmead, editor, bon vivant, indefatigable clipper of odd news items, and a cat person's cat person, had a cat-loving neighbor, an old lady who actually went to the trouble of buying brains from her local butcher and cooking them (sautéed in butter) to provide meals for his aged cat (twenty-four years old!) with no teeth, who naturally required a soft diet. While this surprised some people, it did not surprise us. Margaret often cooks chicken breasts for her cats, boiling them, then serving them up diced, with a generous helping of the warm broth, for her "old guys," and sometimes warmed milk for cold winter nights for the "outside" cats. (The warmed milk seems a less successful culinary treat, since at least one cat loved it, but threw it up afterward.)

Well, perhaps cooking for cats is not so surprising, after all—people will go to extraordinary lengths to provide for their cat (or cats). If that weren't so, how to explain the existence of CatPrin, for example, a custom clothing *tailor* for cats, enabling the proud cat owner to dress his or her cat up in a variety of different costumes, following which you can enjoy the moment, photograph the cat, then, according to CatPrin's instructions, "Remove her clothes and give her a hug."

Judging from the expressions of the cats in the catalog photographs, that would also be the moment when one might expect to receive a serious scratch in return— they do not look like happy campers. Cats, after all, deeply prize their dignity, so being dressed up as a chick, or a frog, or a high-school cheerleader, seems unlikely to bring much in the way of pleasure to a cat, nor do cats have much of a sense of humor, at least about themselves.

Of course clothes for cats are not a more improbable idea than some of the things listed for cat lovers in the upmarket, but relatively sane, Hammacher-Schlemmer catalog. How about a "cat stroller," so you can push your cat out for a healthy fresh-air stroll, like a baby in a pram, with half the stroller made of mesh and the other half of striped fabric, in case the cat wants privacy? How about an English "wicker bunk bed"—hand made by the venerable firm of W. Gadsby & Son, in business since 1864—for cats, so "social cats" can snooze one above the other, after they've worked out who gets the top bunk and who gets the lower one? How about a covered outdoor playground for cats, or a pet door designed to exclude anything but your own cat from entering the house through it? How about a covered basket—Gadsby & Son again, "There'll always be an England"—so you can carry your cat strapped to the handlebars of your bicycle, instead of leaving it at home?

Kittens, of course, are full of playful fun—they chase their own tail, they roll themselves up in a ball, they play hide-and-seek with each other and with humans, but once they're fully grown, cats tend to cultivate an air of pained and reproachful dignity. Not only do they not *do* tricks, they don't much like tricks played on them.

There's nothing a cat likes less than appearing foolish, and most will go to great lengths to avoid it. Indeed this is one of the principal characteristics that cats share with humans. Anybody who lives in proximity to a cat will recognize the way a cat behaves when it has made a graceful leap for a shelf or a ledge and missed—a quick, desperate grapple to pull itself to safety, followed by a furtive look around to see if anybody was watching, then a careful grooming of a paw, as if to indicate that the whole thing went off just as planned. In short, exactly the way humans behave when they have slipped on a wet spot on the floor, or dropped the groceries, or tripped over something: recover, brush oneself off, pick whatever you dropped up, and pretend that what happened was exactly what you meant to do.

The swift recovery of dignity—if at all possible, and at any cost—is one of the salient characteristics we share with felines, as is the instinct for what the French call *l'amour propre,* meaning the need to put the best face on things and to tidy oneself up as quickly one can. You can observe this among humans of the urban persuasion on rainy days in the streets, when a taxi cuts in close to the curb and sends water flying over the pavement from some filthy puddle, spattering pedestrians. Indignation, followed by a quick, furtive attempt to clean up, followed by a quick look in all directions to see if anybody saw (or is, worse yet, laughing), and finally the determined, stately resumption of the walker toward wherever he or she is going, pretending that his trousers (or her stockings) *aren't* wet and splashed with mud or worse. . . .

The animal world is full of perfectly interesting and worthy creatures that *don't* care a damn about this kind of thing—horses roll in the mud until they're filthy, but they aren't a bit of ashamed of it; dogs roll in horseshit and worse and don't feel in the least badly about it; bears don't appear to go in for grooming at all or ever feel the need to look their best, nor do pigs; elephants cover themselves with mud or dirt, the more the merrier; hippos wallow in filth happily—but human beings and cats share the need to look their best in the eyes of others.

This desire not to appear foolish is a strong bond between humans and cats. Cats are in some respects a mirror image of ourselves, and though it's almost always misleading to impute human feelings to animals, cats may be an exception. Dog owners are said to come to resemble their dogs in facial appearance (certainly Michael had a grand-aunt who came to look exactly like her Pekingese), but with cat owners it's the personalities that mesh and come to resemble each other, sometimes so closely that it's hard to tell whether it's the cat that is imitating the person, or vice versa. The late Phyllis Levy, a book and magazine editor with a taste for perfection— even the best was never *quite* good enough for her—was passionately devoted to her cat Tulip. To know Phyllis was to hear all about Tulip, to call her was to hear all the news *du côté de chez* Tulip, who mirrored her owner's svelte grace and unrelenting perfectionism. Phyllis and Tulip faced the world with the same perfect poise and manners, and just as nobody had ever seen Phyllis, to our knowledge, anything but elegantly dressed and made up, Tulip never appeared before guests until she was impeccably groomed, and like her mistress, preferred, when possible, to make a grand entrance—she even seemed to know where to sit in order to be flatteringly high-lit. Her eyes were the same shape and color as Phyllis's, and like Phyllis she could fix her gaze on you in a way that either suggested you were the most important person in the word, or beneath her notice.

Since it was an integral part of Phyllis's profession to eat lunch out—and, very often, dinner—almost every day at fashionable restaurants, where, being Phyllis, she knew the right table to sit at, as well as the names of the owner, the chef, and the maître d'hôtel, all of whom would come over to kiss her hand and tell her how beautiful she looked, Tulip's meals were very often cooked for her by New York's best French chefs, who, knowing Phyllis's devotion to her cat, would prepare special little plats du jour for Phyllis to take home—friandises of minced sweetbreads, or tiny quenelles, or minced game birds, lovingly wrapped in foil shaped into a swan, with a string for easy carrying from one finger. Like Phyllis, Tulip was not a big eater, but had a discriminating palate, and never seemed to put on weight, no matter how many packages Phyllis brought home for her from La Côte Basque, Caravelle, or Lutèce, any more than Phyllis herself did.

Just as it was impossible to imagine Phyllis sitting down to eat a hamburger at her desk—for one thing she liked to wear white and pale pastel colors that don't respond well to ketchup—it was hard to imagine Tulip wolfing down a can of ordinary cat food. Although the fastidious white cat that used to advertise Fancy Feast cat food—its food served in a crystal goblet brought in by a butler—in no way resembled Tulip, every time the commercial ran one thought of Tulip, also waiting, no doubt, to be summoned to a meal by the genteel clink of a silver spoon on fine Waterford crystal.

Both Tulip and Phyllis are gone now, hopefully to a place decorated in perfect taste—Tulip not only fit in with Phyllis's décor, which might have been designed around her, but may have been the only cat in history that never sharpened its claws on the Fortuny fabrics or threw up on the pale peach-colored wall-to-wall carpets—and one still remembers them as bound inseparably together, really a single soul, elegant, refined, never able to pass a mirror without glancing in it, mutually devoted to the simple idea of perfection. Phyllis was about as feline as a person can get while still standing on two legs instead of four, and at moments of delight, she actually seemed to purr. Tulip might easily have been her sister.

When people have a long-term relationship with a cat, what makes it work is often a discrete and perfect match of character and personality, not necessarily the kind of thing you can plan on, or for, but more serendipitous, like the kind of love in which two people fit together like pieces in a jigsaw puzzle, a snuggling, complementary relationship, in which each supplants the deficiencies in the other, and which, when the two are joined together, adds up to more than the whole.

Cats seem very good at reading human character, though they occasionally overreach. Once they're comfortably installed in a house, for example, and have been there long enough to regard it as their own, they may come to assume mistakenly that anyone who enters the house loves cats, and is simply dying to have one sitting in their lap, even though they're wearing a dark blue suit or a black skirt and may, in fact, be allergic to cats. But when they first join the household they're pretty quick to decide who the dominant cat lover is, and what his or her requirements are of a cat.

If what you *want* is a cat on your lap, or on your bed, you will get one—cats have a way of sensing what's required, and most of them are quick to adapt. You want a companion to sit beside you and purr while you watch television? A cat is likely to figure that out pretty quickly. What's more, cats seem able to read human moods. If you're miserable, or sick, or simply feeling the blues, a cat seems to know it, and will appear from wherever it has been sitting to curl up next to you and provide you with warmth, companionship, and—who knows?—sympathy. If you're happy, you're likely to find the cat is happy too, playing around your ankles, and following you around the house with its tail held upright. Whatever is going on in your mind, a cat is likely to sense it and pick up the "vibes."

There are exceptions, of course; some cats, like some people, are insensitive, and not every cat can therefore be expected to play the role of a furry "mood ring" (for those old enough to remember mood rings), but in general cats do respond to their owner's moods, which means that when you're in a bad mood, they'll take refuge under a piece of furniture to ride the storm out, and when you need warmth and spiritual comfort, they will generally do their best to offer it, in their own way. Advice, of course, they can't give, at any rate directly (though if your cats show a marked dislike for a suitor and show it, don't dismiss the possibility that they may be right), but after all you can't expect the impossible from an animal that can't speak.

Of course cats *do* speak, in their own manner; some cats are more vocal than others, but it is usually possible to tell the difference between annoyance, anger, pleasure, indignation, and warning, in the sounds cats make. Even a non-cat lover can distinguish between a contented purr and the screech of an angry or frightened cat when attacked by another cat, and those who are close to cats claim to have a more sophisticated understanding of catspeak. Larry Ashmead goes so far as to claim to have "conversations" with his cat. Well, not just Larry Ashmead, to be fair—a lot of people claim to do this. There's even a Web site devoted to understanding what your cat is saying.

There's no question that conversations with a cat are *therapeutic*—after all, a cat doesn't argue, answer back, or tell you that you're dead wrong, while a human voice, in the right tone, preferably soothing and praising, is something most cats appear to enjoy listening to in moderation. Also, you can whisper your deepest secrets to a cat in the absolute confidence that they're not going to be passed on to anyone, which is more than you can say of humans, including your best friends. Cats probably hear more good gossip than anybody, including gossip columnists, which might explain the self-satisfied expression and general air of superiority that most cats affect.

Most people talk to their cat in their own language, of course, so our cats have always been addressed in English, except for an occasional word or two from Michael in French. Michael remembers receiving many years ago, in his capacity as a book publisher, "one of those bulky manuscripts that looked as if it had been written by a crazy" about talking to cats. This was before Michael had become a cat owner himself. The manuscript had all the familiar signs of mental instability and obsession— the erratic single-spaced typing, done with a faded ribbon, the frayed pages, showing the signs of having been handled by many readers, the innumerable handwritten corrections and additions in microscopic writing, filling the margins like miniature hieroglyphics, in several different colors of ballpoint pen, plus all the evidence of scholarship run amok: glossary, lengthy footnotes, bibliography, source notes, the whole nine yards. Still, as every book publisher knows, many best sellers have been written by madmen. Mere lunacy and obsession have seldom prevented a book from selling in large numbers.

The one thing Michael knew about cat books (learned

from Larry Ashmead, as it happened) was that they always seemed to sell, no matter how strange or unlikely the premise, so he gave the untidy bundle to a couple of editorial assistants, both of whom were cat lovers (you could tell that by the number of photographs of their cats pinned to their bulletin boards and framed on their desks) to read. Both returned deeply impressed, with long, typed reports about how unusual the book was. The author, it appeared, not only *talked* to his cats, but had actually devised a special language which he claimed the cats understood perfectly. He was working now on teaching them to respond in "Catspeak," which was proving to be more difficult than he had anticipated, but he could tell that there were signs of progress, which did not surprise him since he had devised the language based on a careful, scientific study of a cat's ability to vocalize sounds. By the time the book was ready he anticipated being able to appear on the *Tonight Show* with at least one, and possibly two, talking cats. The language, they reported, seemed odd and resembled nothing they had ever seen before, with many different accent marks in improbable places and unfathomable rows of consonants joined together.

Michael took the manuscript home and read it, engrossed until he got to the first examples of "Catspeak," at which point he started to giggle uncontrollably. The language, far from being "invented," was simply Hungarian, admittedly one of the more difficult and unfamiliar of languages, with its strings of consonants, and rows of single and double accents! He recalled the line from *My Fair Lady,* when Professor Higgins's *Mitteleuropäische* rival in linguistics ("oozing charm from every pore") says of Eliza after one dance at the ball that he could tell at once she was, "A Hungarian—and not only Hungarian, but of royal blood!"

Of course anybody pretending to invent a language for cats might easily suppose that few, if any, readers at a New York publishing house would recognize Hungarian—it certainly has the appearance of a language not intended for humans, or at any rate Anglo-Saxon humans—and it would be nice to suppose that the book was eventually published, and that somewhere out there are well-intentioned cat owners patiently reciting Hungarian to their cats, in the forlorn hope of getting a reply. Or perhaps they *do* get a reply. You never know with cats!

(Larry Ashmead had a dog that buried his wallet somewhere in the garden. Unable to find it, he contacted a woman in Los Angeles who claimed to be able to speak to dogs. She had several conversations in Hungarian with the dog over the telephone, but the dog never revealed to her where the wallet was buried, nor did she reveal why Hungarian was the language of choice for dogs.)

In any event, Margaret certainly seemed to communicate with Irving, and had long heart-to-heart conversations with him in a low, soothing voice. After all, if it's possible to be a "horse whisperer," why not a "cat whisperer"? One thing you can say for cats over horses—nobody ever suffered a broken bone because a cat stepped on their foot.

One of the things Irving communicated was the strong desire not to have another cat in the house. Irving had been—whoever Margaret's husband might be—the one and only cat of the house, master of all he surveyed. Nobody pushed him away from his feed bowl, or shared his litter box, let alone his place on the bed, where a fur coverlet was his favorite place for napping and sharpening his claws.

Once, when we were going away for a vacation, we boarded him at an establishment on the West Side that was a kind of storefront urban summer camp for cats, where they lived in what looked like a miniature-size child's playground and were encouraged to "interact."

Irving had not looked happy at all when he was left there, and so it proved, since when we got back the two stout ladies who owned the place asked Margaret not to bring him back again. He had not, they complained, been willing to "join in" and "participate" with the other cats, or play with them—indeed he had spent most of his time hiding under a piece of cat furniture, drooling and looking miserable. He had not adapted to "the spirit" of the place, rather like a child sent to camp who refuses to engage in team sports.

One problem may have been that having been around Margaret since earliest kittenhood, Irving may have thought of himself as a person, rather than a cat—in any case, no further attempts were made to "board" him. He was our only cat, city or country, until his death, which is just the way he would have wanted it (well, he wasn't really "ours," he was hers). His death, when it came, was a traumatic moment for Margaret, who had come to think of Irving as a close and beloved friend, for some period of time the most stable and dependable part of her life. Unlike a husband, he did not become ill-tempered or make unreasonable demands; unlike a human friend he did not take to drink, or give unwelcome advice, or age suddenly and ungracefully. He was there for her whenever she needed warmth, companionship, and unquestioning love, which is not a small thing. At first, once Irving had passed, Margaret was reluctant to contemplate having another cat—no cat, she felt, could replace Irving—and resisted suggestions to look for a nice kitten. Still, Margaret is somebody for whom a total absence of cats is unthinkable and unbearable. Inevitably, with a certain amount of guilt, her thoughts eventually turned toward another cat—not exactly a "replacement," since no cat could completely replace Irving in her life—but perhaps a *substitute*.

At that time we were living on Central Park West, overlooking Central Park, and our downstairs neighbors, Phyllis and Stanley Getzler, had become friends, in part through their ownership of a very friendly golden retriever named Missy, whom Margaret liked to take for walks. It was love at first sight between Missy and Margaret from the moment they met in the elevator, and instant friendship followed between us and the Getzlers as well, once we had been introduced by the dog. Phyllis, though not a cat person, had met Irving, and felt strongly that Margaret should have another cat as soon as possible, and that any animal would benefit from being owned by Margaret, which is true enough, and so joined her in the search. Phyllis Getzler was—and remains—a woman of strong and forceful personality, so she quickly pushed Margaret into action, and joined her in the cat hunt. Margaret was looking for an orange male kitten to replace Irving.

Michael paid very little attention to this hunt, and was therefore surprised to get a telephone call from Margaret while he was at his office, announcing that she and Phyllis had found a cat at the New York City animal shelter on the Upper East Side, or, to be more exact, *the* cat. It was everything she had been looking for, Margaret said on a pay phone from the shelter, young, very pretty, friendly. . . .

That sounded terrific, Michael said. What color was it? Well, it wasn't orange, he was surprised to hear. It was gray-and-silver striped, Margaret said, with black markings, and big blue eyes, really *very* attractive. . . . There were just a few little problems, but nothing we couldn't learn to live with. These were, it turned out, that the cat was not a kitten, not a male, that it was full grown, that it had a suspicious large lump in the belly that might be a hernia or a tumor, and that it was missing one foot.

Missing one *foot*? This was not the result of an accident, it turned out; the cat had been born with a deformed left front leg, shaped rather like a miniature seal's flipper, with a useless paw at one end. You wouldn't ever guess it, to see her walk, Margaret enthused—she really handled her deformity very well, and it hardly spoiled her looks at all. . . . Michael hung up wondering how it was possible in a place where there were dozens of four-legged cats begging to be adopted, to pick out one with a hernia or a tumor and three legs.

That she handled her deformity very well became evident when Margaret and Phyllis brought the cat back to our apartment building, in a cardboard cat carrier provided by the shelter, which bore on its side a stencil reading, "I am going home!" The bottom dropped out of the carrier in the garage as it was being lifted from the seat, and the cat vanished into the depths of the garage at a remarkable speed for a three-legged animal, or even for one with four legs. Her coloring might have been designed to aid concealment in a New York City garage, with ramps leading down to two more floors of parked cars, dimly lit from above, and hundreds of concrete nooks and crannies in which to hide. Margaret and Phyllis—soon joined by a couple of the parking attendants—went from car to car, getting down to look under each one, and making encouraging noises, but as everybody knows, a frightened cat's instinct is to stay put and avoid contact with strangers. There was always the danger, too, that the cat, if cornered, might make a dash up the ramps and out into the street, where it was only too likely to be hit by a car. In these circumstances, as every cat lover knows, there is nothing for it but to proceed with caution and hope for the best, and the cat was eventually discovered, and brought up to the apartment in Margaret's arms, apparently undamaged by the experience.

By the time Michael got home, the cat had not only settled in, but been given a name, Queenie, after the title (and the heroine) of Michael's novel about his aunt, the late Merle Oberon. Like "Auntie Merle" (whose own real name had been Queenie, before she adopted a more glamorous name for her film career) Queenie's beauty—for despite the stunted leg, she was a very handsome cat—concealed an impatient, imperial, and dictatorial nature. She did not take interfering with lightly, nor having her wishes ignored. When annoyed or challenged she gave the offender a sharp whack with her flipper, a kind of warning slap.

God knows where she had been, or how she had ended up at the animal shelter, but she settled into life at 55 Central Park West as if to the manor born, and sat long hours at the big picture windows, looking down at Central Park. As it transpired, a good deal of her life would be spent looking out of windows, even in the country, since outside she could neither defend herself, nor climb a tree, so it was dangerous to let her out. Queenie had to content herself with sitting by a window or on a screened-in porch, staring malevolently out at the birds with her stunning Paul Newman–blue eyes as they flashed by or landed in the carved stone birdbath.

No doubt she imagined herself reaching up and swiping one of them out of the sky in a cloud of feathers, but it was not to be. Her tiny front "flipper" did not inconvenience her at all, and she had developed a whole lot of ways of hiding it, so that only the most observant of cat watchers ever noticed anything wrong with her.

Queenie was, in many respects, the perfect cat. She would curl up happily in your lap while you read, or worked, or napped, or doze in a patch of sunlight while you went about your business. If she was in any way displeased, or felt that her feeding time had been delayed beyond reason, she would stomp across and slap you with her stunted front leg to remind you of your duty. She not only had personality, she achieved a modest degree of fame by appearing in *People* magazine, but she did not let it go to her head.

Though Margaret's initial comment was that Queenie "was no Irving," she stepped into the role of number one (and only) cat without hesitation. She was not quite as loving as Irving had been toward Margaret, or nearly as gentle, but she did her best, shuttling back and forth between Poughkeepsie and New York City every weekend in the back of the car like a pro, though with a tendency to use her litter tray on the floor in the back the moment the doors and windows were closed and the car was in motion. Otherwise, she was a pretty steady traveler, quite capable of going to sleep on the shelf beneath the rear window of the car. You just had to remember that she could move pretty fast for a cat with three legs, and be careful when you stopped at a tollbooth—there was always the nightmare thought that she might go leaping out the window into the traffic on the Henry Hudson Parkway, and use up all nine of her lives in one go!

As a house cat she liked company, but retained a certain aloof quality. Certain places were hers, and she disliked being evicted from them by, or for, visitors. Like most cats, she was an instinctive tactician, and thus had a marked preference for the high ground—the top of the refrigerator, for example, which placed her slightly above the level of a human head, and out of reach of vacuum cleaners, mops, the very rare visiting small dogs, strangers, and plumbers. Also, as you would expect for a creature born with a deformed limb, she had a perfected sense of strategy—she instinctively chose a position with a good view, an escape route, and some degree of camouflage. Animals that have been abused—and, alas, so many that end up in animal shelters have—almost invariably show some signs of it, sometimes nervousness, sometimes sudden aggression, as if they expect at any moment to be physically punished or attacked, but Queenie had a degree of self-possession that would have been remarkable in a person, let alone a cat, together with an imperious nature that brooked no unwelcome familiarity. She didn't look starved, either—she was as glossy and sleek as could be, right from the shelter. Why, one wondered, had her owners gotten rid of her? Was it the stunted little flipper of a front leg, which we felt actually added to her charm and character? Did she not get along with children (we never had occasion to put this to the test)? Or had she been a touch too aggressive in enforcing her wishes? Who knows? Animals can't talk, so we never really learn their stories about the past. We can see the physical effects of abuse and neglect easily enough, but absent these, who knows what a cat's previous home was like, and whether it was well treated, let alone whether it misses its previous owners or not? The cat certainly isn't going to tell us.

Most cat owners tend to assume that the previous owners of any cat from the shelter, or that turns up at the door looking for a new home, were beasts in human form, but who knows if it's true, let alone if that's what the cat thinks? In any case, once Queenie had settled in it was as if she had always lived with us, and it was hard to imagine waking up in the morning without seeing those bright blue eyes fixed on you, staring out of a silver, cream, and black-striped face, and wondering how on earth anybody could get rid of her. One happy result of her deformity was that she couldn't really scratch at carpets or furniture, which made her the perfect cat for two people who were growing broke furnishing a house in the country room by room. (And this was in the days *before* we had a decorator, when we still drove down to Paramus to buy furniture on the cheap, hauling it home in the back of our Chevy Blazer, instead of meeting Thom von Buelow for dinner in the Grill Room of The Four Seasons, and picking antique furniture and rare—

and expensive—fabrics out of the apparently bottomless Vuitton bag full of swatches, sketches, and photographs he carried with him.) Whatever other faults Queenie may have had, you could at least be sure that she wouldn't tear your cut velvet sofas to shreds, though she might, if she felt so inclined, throw up on them from time to time. You were never woken up by the sound of a cat trying to tear a mouse-size hole in your beloved oriental carpet, or sharpening its claws on your favorite needlework cushion. No doubt had Queenie been able to indulge herself in this kind of destruction, she would have been happy to, but she wasn't.

As we got more cats, Queenie watched them attack the furnishings with a good deal of interest, and no visible sense of outrage or disapproval, but in the meantime, there was a blessed period in which it actually seemed that we might be able to have the kind of elegant house that so many of our friends had, in which you weren't always covering things with a blanket or a shawl to hide a hole in the fabric, and in which the backs of the chairs weren't torn to shreds, exposing the innards. That was not, of course, going to last, though hope springs eternal, leading one to look backward and ask how in the world we ever supposed it was a good idea to have this chair or that sofa upholstered in magnificent cut velvet at God only knows how much a yard when it was always clear that we were not just going to have a cat, but *cats*, and that by any reasonable standard of statistical probability, most of them would have two intact front legs and, therefore, two working sets of claws.

Of course some people solve this kind of problem by having their cats declawed, but apart from the fact that it seems like a cruel, unnatural, and painful procedure, you can't really put a cat without claws out the door in the country, where climbing up trees, gutters, and fence posts and fighting back, as it were, "tooth and nail," are all part of the cat's basic survival kit. It's like keeping an *unloaded* pistol around the house—it may seem like a sane and safe idea to non-gun people, but gun owners take the view that in the one brief and terrible moment when you might actually *need* a gun, it had damned well better be loaded or it's no use, and also that the only way to be safe around guns is to treat every gun as if it *were* loaded, no matter who tells you it isn't.

The cat's claws are its ultimate weapon, and the reason it spends such a lot of time keeping them sharp isn't just to destroy your furniture and carpets but because if and when the cat *needs* those claws, they'd better *be* sharp. This is nothing cats need to be taught—it's part of their genetic programming. However peaceful a cat may look dozing on your best chair, no cat is a pacifist, or in favor of disarmament—cats are predators, born to kill, vicious and determined in defense of their territory, and not afraid to draw blood. Like people, of course, some of them are more timid than others, while some are born bold, and like children in a boarding school a lot of their play has a serious purpose, which is to figure out where other cats in the immediate vicinity belong on the scale of aggression. Nice as cats may be toward humans, they can be pretty brutal toward each other, and are capable of inflicting fairly severe wounds when maintaining—or, more dangerous, challenging—their position in the "pecking order," or defending their territory against a newcomer. Queenie was by no means timid, but her weapons of attack were limited by that useless foot. On the other hand, like a lot of cats, she had a real gift for wedging herself into a corner and puffing her fur up until she looked about twice her usual size—that, combined with a snarl that exposed her sharp fangs and a low, menacing growl, was her best and most convincing defense. She could make herself look pretty terrifying when she felt it was called for, unlike poor Irving, who had been the most peaceable of cats, without, so far as one could tell, any impulse to fight—"a quiet life," might have been his motto, and by and large that was how he had led it.

Queenie, on the other hand, had a wilder streak, and was not necessarily a couch potato. She expected to get her way, and by and large she did, all the more so since she was the only cat in the house, and remained so for a long time. The occasional sight of another cat out the window did not make it seem as if she would welcome a companion. She hissed, spat, snarled, and pressed her nose up against the windowpane, as if sending the message that she was ready and waiting to defend her turf.

Outdoor cats, hardened to rejection, were not much impressed by this. From the very beginning, one of the first things that struck us, in fact, about our house in the country was that it seemed like a magnet for cats— equally astonishing was the sheer *number* of cats out there in the landscape, a never-ending supply of pitiful, or defiant, or hardy, survivors. Some of them became familiar, some of them simply seemed to be passing through on their way to somewhere else. Needless to say, Margaret put out food for them, which only increased the number of mostly nocturnal feline visitors, not to speak of attracting possums, raccoons, and skunks.

The amazing thing was not just that there were so *many* cats out there, but that so many of them appeared to survive somehow, despite the harsh winters—yet they were clearly domestic cats, not wild ones, that had once lounged on somebody's sofa, or dozed in front of the fire. At some point, it was clear, somebody had lost patience with them, or perhaps they had failed to get along with a new baby or boyfriend, and out they went, probably carried off by car in a cardboard box far enough from home so they wouldn't come back, then dumped unceremoniously by the side of the road or in the woods, good-bye and good riddance.

Or, who knows, people moved from a house to a small apartment in which pets weren't allowed, or moved in with somebody who was allergic to cats, or just decided that having a cat wasn't what they had really wanted in the first place. . . . For somebody like Margaret, much of whose life was devoted to the safety and the well-being of her animals, it was hard to understand how people could simply get rid of an animal that had once sat on their lap, or snoozed at the foot of the bed, that had, as it were, paid its dues. But from farm to single-width trailer on cinder blocks, all over our neck of the woods there were people who didn't suffer from the same compunction, and as word went out among local felines that there was always a bowl of food on the porch of the old Hubner house, the number of eyes gleaming in the woods as night fell and the porch lights went on seemed to multiply week by week.

Sometimes we would see these stray cats by day too, though less often. Some of them were not exactly "stray" either, in the full meaning of the word. For better or worse, they had made our farm their territory and their hunting range. They seldom approached the house by day—no doubt they had learned you were only too likely to get an angry shout warning you off, at best, or a kick or a well-directed small stone at worst, if you got too close to a strange house—but since we rode every day, exploring our land (and our neighbors') on horseback, we got used to seeing the same cats out in the fields several times a week. Most of the them would run off when they saw the horses approaching, some of them would simply sit and stare back at us indignantly, as if we were trespassers, and one or two hid, peeping out from the leaves and the high grass with a mute look of appeal.

One of the shyest was a thin, frightened cat, gender unknown, orange with white markings, no longer young, certainly not a kitten, and with the unhappy look of a cat who had never expected to have to forage for food in the great outdoors.

Chutney, Margaret called him, after his color, and was determined to bring him into the house.

4. Chutney

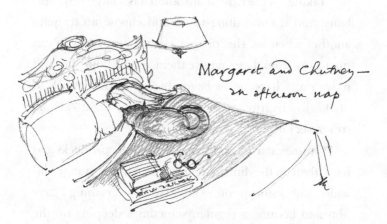

Margaret and Chutney — an afternoon nap

O ver the years," remembers Margaret, "they have arrived at different porch doors, the laundry room door, in the barn, emerging out of culverts, hiding under jumps, in the weeds, up from ditches. Some visit from time to time, never to stay, others settle in quickly. Some have been alone, others in pairs, or with families, fat, starving, sick and well, old, young, friendly, feral, shy, skittish, all of them abused or abandoned. Some have had to be put to sleep. Few make a total recovery— anything abandoned rarely does. Old fears die hard. It may take years. It may take forever. Most stay. A handful find wonderful homes—we would never let them go otherwise.

"Taking on the care of any animal is a large responsibility, and it's one thing to go and choose a cat, quite another when he, she, or they choose you. Oh, you can turn away from them, ignore them, or just hope they will move on. I have never been able to do that, much to Michael's chagrin. At this time we have eight cats, all strays, six indoors.

"Chutney started appearing around the paddocks and barn, then on the dining-room porch. He was orange and white, with a stocky, round face. He ate everything I gave him and became a regular, sometimes sleeping on the porch, in a cardboard box lined with an old saddle pad and covered with one of Michael's old goose-down Eddie Bauer jackets. 'It'll have to be thrown out now,' Michael said. 'He's probably full of fleas, you don't know anything about him, he looks filthy, and he's not fixed. He could even have mange.'

"We were away for three weeks soon after, and although I left his care in the hands of Dot Burnett, who was our housekeeper then, she did not see him. Two days after getting home, he showed up again, looking even more thin and dirty. 'He's back,' I said to Michael, 'and for good, this time.'

"I set up an overnight for him in one of the guest bathrooms, carried him up there, and closed the door. He was not happy and cried all night. 'You see,' Michael said, 'he hates it, he doesn't want to be here, and God knows what he has brought into the house.'

"'I'll take him to the vet in the morning,' I said. 'The first appointment I can get.'

"It's an old scenario by now. I may house them in different locations for a while, but they mostly end up staying with us, although a few we have given away to friends. Chutney was a loving, quiet cat, and I wonder often how he ever survived for as long as he did in the wild. He had the odd habit of often peeing in the kitchen sink. Michael never knew for a long time, and when he did find out was, as I expected, pissed off. I thought it quite smart and very easy to clean. Chutney did not make friends easily with other cats, but for years he slept against my legs."

Actually, it was more than that. Queenie, with her independence and bossy ways, had never quite replaced Irving in Margaret's mind, fond as she was of her, but Chutney was in many ways like a reincarnation of Irving. With his long whiskers and his vivid coloring and his big, round head, he was an impressive cat, but a real gentleman, well-mannered, quiet, and utterly devoted to Margaret. He was more adventurous than poor Irving, and quite enjoyed a leisurely stroll in the great outdoors, what elderly German and Austrian gentlemen call a *Spaziergang*, whether by day or in the evening. Snow and rain or heat and humidity did not discourage him from his daily walk—after all, having lived as a stray for so long, the outdoors did not frighten him, though he preferred to stay within sight of the house, so he could cut a hasty retreat if necessary to the porch.

To say that Chutney was not aggressive is putting it mildly—he did not attack other cats, he did not bite or scratch people, he didn't even sharpen his claws on the furniture as a way of starting his day. He avoided fights, and therefore never came home with the kind of wounds that mean making an emergency call to the vet.

That does not mean his life was stress-free, of course. No sooner had he been released from the upstairs bathroom while he recovered from being neutered and allowed to wander all over the house than Queenie, who was at first anything but pleased at the arrival of a stranger, bit Chutney so badly that his leg abscessed and he almost died. He *did* have to be carted off to the vet again, and on his return hid for weeks behind the furniture, out of Queenie's way, making a slow recovery. It's amazing how cats survive at all, given the effect that one cat bite can have (after all, they are meat eaters who don't brush their teeth), in terms of damage and infection. Poor Chutney looked like a goner for some time, but eventually massive doses of antibiotics and his native good health (or good luck) brought him round, though for a long time he sensibly kept his distance from Queenie.

Then, as is so often the case, they settled down and became friends, the pecking order having been established. Chutney, to give him his due, didn't seem to care who was "top cat," and was happy to leave all that to Queenie. He himself had no ambition except to sleep as close as he could possibly get to Margaret at night, and, if possible, to share a late afternoon nap with her. He had, quite possibly, the kindest face it is possible to imagine a cat having, broad, good-natured, trusting, and even while making his rounds of the farm he showed no inclination to attack small birds or baby rabbits or moles, or any of the creatures that cats usually try to torment, then kill if they cross their path—if there is such a thing as a pacifist cat, it was Chutney.

He was also, of course, the first stray cat we adopted, after some difficulty, for with strays that is usually the case. Short of running around with a net, or more practically speaking, a Havahart trap, it's hard to pick up stray cats if they don't want to be picked up, and very often they don't. Even when they *do* want to be picked up, they do not necessarily make it easy. Chutney had apparently seen enough of us—of Margaret, at any rate—to conclude that we were likely to be friendly, had taken his time, months of it, to case the house, and had clearly had enough of living in the great outdoors on his own. It's scary out there for a little cat, with temperatures that go down to ten or twenty below zero at night in the winter, and coyotes that hunt in packs, or in the heat of the summer, bitten by fleas and ticks. Cats are sly creatures, adept at cutting a low profile, and at hiding in places that don't seem big enough for them to get into in the first place, but your average house cat, while it has built-in survival skills, isn't really cut out for life in the wild, and doesn't necessarily enjoy it. Most of them have simply been thrown out to fend for themselves, and along the way have developed—with good reason—a deep, abiding distrust of human beings that is hard for even the most determined cat lover to overcome. Having been thrown out of its home, the cat's attitude in most cases is, not expectedly, "Once bitten, twice shy."

Margaret pursued "Chutney" (who was not yet named, of course), with strategically placed cans of food, drawing him nearer and nearer to the house, and finally onto the porch itself, but however hungry a cat may be, this doesn't always work, nor does it always lead to the critical moment of reaching out to *touch* the cat. A lot of cats will leap away at that, and may not come back for a long time, if ever. They are only too likely to equate an out-stretched hand with being hit, and react accordingly. What is required, basically, is a huge amount of patience—the patience to try the same thing over and over again, night after night, until the cat begins to build up a slight willingness to take a risk, enough confidence to come close and let itself be touched, at which point, with even more patience, it might just let itself be drawn inside for a moment or two, just to scent out if there are unseen dangers in the house, or perhaps to get a taste of what it's like to have its feet on warm carpet again, instead of cold snow.

That Chutney *wanted* to come in out of the cold is certain—you only had to look at him to see that—but it took a good deal of effort for Margaret to gain his trust and confidence to some degree, and a good deal more to get him upstairs and locked into a guest bathroom where he must surely have felt like a prisoner before he finally settled down for a nap on the bath mat, and resumed life as a pet, as if it had never been interrupted.

These are the victories that matter—and are never forgotten—in bringing stray cats indoors. Once Chutney came, survived a trip to the vet, then Queenie's attack, and as a result of it, another trip to the vet, and finally settled in, he never showed any desire to stray farther than the flower garden outside the front door, within easy reach (and view) of the house, and he quickly became such a fixture of our lives that it was hard to remember that we had once merely caught glimpses of him from a distance while we were out riding.

Not every cat, as we were soon to discover, makes the return trip to domesticity as easily and successfully as Chutney did, or with such good long-term results. Nor, of course, once you start rescuing strays, can you *keep* all of them, unless you want to end up as one of those crazy cat people who are occasionally found by the police or the ASPCA living in a tiny house with forty or fifty cats.

There's some point at which having a lot of cats crosses the line from normal behavior to insane, but it's hard to fix a firm number. Neighbors of ours, whom you will meet, have anywhere between a dozen and twenty, depending on the circumstances, and while that doesn't necessarily make dinner at their house a treat—you are likely to have not just a cat on your lap, but cats walking across the dinner table, drinking out of your water glass, and snatching food off your plate or fork—they have the

situation under some kind of control, and the cats are lavishly looked after, well fed, and obviously loved as individuals. However, forty to fifty cats in a small house, particularly, which is often the case, owned by an old person who can no longer look after them properly, is of course a whole different story.

From the very beginning, we have tried to keep some kind of distinction (though not very successfully) between those parts of our house that are zoned for cats, and those parts that aren't, unlike certain of our neighbors who simply threw in the towel and abandoned their house to their cats altogether. This has meant, among other compromises, giving up one bathroom and the downstairs laundry room to the cats, establishing the formal living room, which admittedly we seldom use, as a cat-free zone (there's a glassed-in door that allows them to peer in, but not to enter), and keeping the cats out of the guest rooms as much as possible, though it's up to guests, who are few and far between, as to whether they want to close their door at night or not. If they do not, they are likely to find a cat on their bed, or sometimes on their pillow, when they wake up, which is likely to be early, since our cats are used to having their breakfast between six-thirty and seven, and mildly impatient when any human being stays in bed later than that.

Still, at least we *warn* guests, and leave it up to them whether to close their door or not. In the days when one remembers having been a guest oneself in country houses, one was often woken in the middle of the night with a start when a previously unseen cat launched itself onto the bed in a single bound, landing with the impact of a Scud missile, claws extended. Etiquette seems to demand that guests be fully aware of the cats, and allowed to make up their own mind as to whether they want to share the bed with them. Guests bringing dogs are, naturally, asked to keep their door closed, in order to avoid a protracted cat-and-dog fight in the middle of the night, ending either with a visit to the twenty-four-hour-a-day emergency animal hospital off Route 55, or the total destruction of what little remains of our carpets and upholstery.

Actually, the cats seem to accept the presence of dogs with better manners than they would a visiting cat. They seem to be able to figure out that the dogs are under strict control, and forced to remain on their best behavior. Their presence in the house doesn't exactly *thrill* the cats, but since they're of a different species, it also doesn't threaten to upset the pecking order, either, as another cat's presence, even behind a closed door, most certainly would. The knowledge that there are dogs in the house brings the cats together, in fact, in a close, huddled bunch, like seamen preparing to repel boarders, their fur raised and puffed out until they look twice their normal size.

Of course none of this was a problem when we had only two cats and many guests. Now, twenty years later, we have lots of cats and hardly any guests at all, so it's only natural that the cats have taken over a good deal more of the house than used to be the case, or that they feel that the house basically belongs to them. Back then, when it was just Queenie and Chutney, guests either made a fuss of them, or didn't, and the two cats mostly stayed out of the way, or settled in a comfortable position on a chair or sofa to watch the proceedings from a distance. Of the two, Queenie was by far the more outgoing and, when in the mood, "pushy," whereas Chutney preferred to keep to the lowest of low profiles, in extreme cases hiding underneath a bed until Monday morning, when the guests had all gone back to the city and he felt it was safe to come out again.

We had one perennial guest, Mayo, a friend whose noisy ways, which included dropping glasses, doing her yoga so violently that she cracked the ceiling of the room below hers, and a tendency, after a few drinks, to knock things over by accident and argue at the top of her voice about politics, used to send Chutney in search of a place to hide in peace and quiet at the first sight of her in the driveway—all the sadder, since it was among Mayo's many illusions that she had a natural rapport with animals of all kinds, and that they loved being in her company. Chutney, who disliked smoking, loud noises, arguments, raised voices, or any other disturbance, needless to say did not share this particular illusion.

For many years he filled a role in Margaret's life that was special, and knew it—he was her emotional stabilizer, her adoring companion, purring away like a well-tuned motorboat as he lay at her feet, or on her lap, or cuddled up next to her on the bed. His idea of heaven was a long nap with her on a cold day, always a silent, unconditionally loving presence, never about to spoil the mood, as so many cats do, with a sudden swipe of the claws. He was never bored, always glad to see her, and when she came home, he could often be seen at a window near the front door, waiting for her faithfully, his shiny yellow eyes fixed on the pathway of uneven, irregular paving stones that led to the front door. He was too dignified to do the equivalent of shouting out "You're home! You're home!" by jumping up and down like a dog—his expression was forgiving, as one might forgive a daughter who had stayed out too late, and relieved, as if he was thinking, "Who knows what she gets up to when she's out of my sight?," and perhaps just a touch reproachful. He had that element which is almost as hard to find among cats as it is among humans: kindness.

Chutney was, in fact, a prince among strays, so much so that it was hard to remember that he had once roamed wild in the fields around the house. Not every stray turned out to be that successful at adapting to domestic life, of course. Some had to be moved on to other homes, and some, as you will see, never made the grade—the saddest stories of all. But Chutney not only made the grade, he defined it. Queenie could be temperamental, often ate too fast and threw up on the carpets (in those days, we still worried about such things and came running day or night, with soda water—said to be good for dabbing off stains—paper towels, a spatula, and, if worse came to worst, a spray bottle of Resolve, with its endless list of warnings about the ways in which it might fade or bleach out colors if you didn't follow the directions *exactly,* all of it printed in microscopic type which you weren't likely to read at three in the morning on your hands and knees in the freezing cold, after Queenie had had second thoughts about her dinner), but Chutney seldom if ever made a mess, or prompted us to ask ourselves what we had been thinking of when we were persuaded to spend such a lot of money on pale green (*pale green,* perfectly calculated to show every stain and spot, including the circular bleached ones made by overenthusiastic use of Resolve!) wall-to-wall carpeting, and beautiful rugs. Chutney could have lived in a shop full of priceless oriental carpets without risk to the merchandise. Really, he had no bad habits, unless you count a total, loving dependence on Margaret.

Of course you can never tell with strays. Some work out, some don't, there's no way to know in advance. You learn something from each of them. It's a whole different thing from that of buying a kitten in an upmarket pet shop—you're getting a cat with a whole lot of prior issues and ingrained habits, one that has survived rejection, starvation, freezing cold, and brings with it everything from ticks, fleas, and neglect to serious illnesses and a major chip on the shoulder. A few, like Chutney, gradually turn themselves into ideal cats, a few, more rarely, don't even require a period of transition—they walk through the front door, groom themselves, take a walking tour of the house, and settle in as if they had always lived here. Chutney in his time would see a good deal of *va et vient,* the comings and goings of numerous cats, some successfully adopted, some, alas, not.

At a much later date there was Chui, for example. As Margaret says, "'Chui' is Swahili for leopard. I called him that because his coloring was similar to a leopard's. He was around nine or ten months old when he sauntered down the path one day, his right ear folded over. He only stayed with us a short time, as he had one unfortunate habit of peeing on the other cats when they got into a tussle. It placed him in a no-win situation. But we found him a home with a dog, who quickly put an end to his habit, and they are living happily together."

Poor Chui was a perfect example of a stray cat with a problem you couldn't have guessed at in a million years. Not only couldn't we have guessed at it, neither we nor the other cats could believe it when it happened. The irony of it was that he was in every other way an attractive and well-behaved cat, except that his secret weapon was surprise—coupled with the sheer shock the other cats felt when they got peed on. They virtually *quivered* with indignation! A slashing tooth-and-nail attack they might have taken in their stride, but Chui not only caught them by surprise, he deeply humiliated them. They did not forgive him, and neither did we, since he generally attacked them when they were gathered on one of the nicer couches, and the stain and odor of cat piss tends to stay around for a very long time indeed.

Then there was Mrs. Bumble, named after the wife of the beadle in Dickens's *Oliver Twist*. Margaret describes this particular stray thus: "From ditch to dining room in three weeks! That's all it took Mrs. Bumble. She was a long-haired tabby with a magnificent tail, like a grand plume. She went through all her blood tests with flying colors, had her shots, and was spayed. She arrived plump for a change and took great care of her appearance. Got along well with all the other cats. Then one evening, as I was going down the stairs, I put my arm through the banister railings to stroke her. A mistake. She grabbed my arm and bit me from hand to elbow. Shit, I thought, I should go to the emergency room at Vassar Hospital, but I could be there for hours, and it's the women's finals at the Australian Open on tape, I can't miss *that* . . . so I cleaned it up, drenching the arm in iodine, and wrapped it. By the next evening I had put myself on antibiotics and two days later I was in the office of a hand surgeon. He was gloomy and thought he should open up all the puncture wounds. I bargained for another twenty-four hours with antibiotics, and to take it from there. Within a month she had bitten me again, no warning, just purring away in my lap. "It's your hand again this time," Michael said. "It could have been your face, or an eye." He took her to the vet's early the next morning."

It's not often that you get a cat that's unpredictably *dangerous*. You learn pretty quickly that some of them may bite in certain conditions, but you also—if you have any smarts at all—learn equally quickly to avoid those conditions. Most cats, in any case, give ample warning of their intentions. They give a low, deep-throated growl, or puff themselves up like old-fashioned muffs, every hair on end, or signify in some other way their intention to attack. If you persist in what you're doing—grooming them with a stiff brush or a metal comb when they don't want to be groomed, for instance—then you can hardly complain when they try to bite or scratch. Mrs. Bumble was in a different category altogether, poor thing—she was a "stealth cat" if ever there was one, sitting there as peaceful as can be, and striking out all of a sudden and for no understandable reason. Margaret might have patted her two dozen times, with no consequences at all—in fact with every sign of enjoyment on Mrs. Bumble's part—but on the twenty-fifth time, out of the blue, she sank her fangs in, again and again. A mental imbalance? Some trauma from her previous experience as a house cat? Who knows, who can say?

We debated a long time before reaching our decision, but there was no alternative, really. When she was in a good mood, Mrs. Bumble was the nicest and most affectionate of cats, and she was certainly attractive—a really striking cat. She was a tough survivor too, having spent a few weeks of the winter in a drainpipe to one side of our driveway. Still, what are you going to do? You can't give away a cat that bites without provocation, and even our vet Mike Murphy, who is often (surely against his better judgment) drawn into attempts to get a cat adopted rather than put down, had to admit that he didn't want the responsibility of putting Mrs. Bumble in somebody's home only to have her turn on her new owner. There's always the possibility of the outright lie—"Oh, she's a dear, wouldn't hurt a fly, you'll love her!"—but that doesn't seem fair, and when the worst has happened, and somebody else is in the emergency room, how is one going to feel about that?

So Mrs. Bumble had to be put down—there wasn't really a choice—a glum day for all concerned. It's hard enough to put down a cat when it's old, or ill, or in pain, but much harder when it's plump, healthy, and glossy, brimming over with élan vital, as Mrs. Bumble was. Even Mike Murphy—a man whom one might have supposed had become used to animal tragedies—looked at her doubtfully. It might have been useful if Mrs. Bumble had chosen to attack him at that very moment, sitting on the examination table, but no such luck. The ladies at the animal hospital looked at Michael as if he were Heinrich Himmler, when he left her there, but then they hadn't gone to the emergency room with Margaret, or seen what her hand and arm looked like after one of Mrs. Bumble's surprise attacks. Not all stories have a happy ending, and that's all there was to it. Having said this, over nearly twenty-five years of feeding God only knows how many strays, tempting them close to the house or even into the house, adopting them, or getting them adopted by people who were looking for a cat, Mrs.

Bumble was the only one who turned out badly, and could neither be adopted nor given away.

Chutney saw some of them come and go, liked some of them, had his doubts about others, but retained his position, not exactly as First Cat—he wasn't about to fight for a position on the pecking order he didn't care about—but as senior, and in Margaret's case, most beloved, of cats. He had a kind of Buddha-like wisdom and calm, which we often admired and envied, and an ability to take the rough with the smooth in life which we might all have done well to emulate. He seemed like a permanent institution, and it was hard to imagine the house or the farm without him. And so it was even more of a shock when, quiet to the very end, he passed away. "One snowy afternoon," Margaret remembers, "he went out for a walk and when it got dark and he had not come back, I went out looking for him, and found him lying on a garden path. He had not been dead for long. I sat in the snow and picked him up. 'You're all right now, you're with me,' I said. We had him cremated and his ashes were in one of my closets, along with his bowl, for many years."

5. Mumsie and "The Terrible Twins"

Mumsie appears with her family

After Chutney, we never looked back, or, to be more exact, we never looked for cats. They, on the contrary, looked for *us*. It wasn't a question of *finding* a cat, cats simply appeared, "out of the woodwork," as the saying goes, and it was more a question of trying to decide which ones should be invited in. A preliminary test period was usually called for, which in some cases was protracted.

Margaret didn't take her time making her mind up about Hooligan, however. "Hooligan," she remembers, "was abandoned with two other cats in a house just up the road from us. The tenants moved out overnight in a bitter February, leaving doors swinging open. Along with a young woman who was working for us then, I went up to take a look inside, and found one room with several empty cans of cat food. Immediately, two cats appeared, and a third stayed crying outside. We managed to get hold of the two, but never the third. I kept the two of them in our barn laundry room with food and water for several days without letting them out. 'Hooligan,' as we called her, was jet black with lime green eyes, everything I would not choose, other things being equal, in a cat, so long-haired that to this day she often takes on a spiky look, like that of a punk rocker. She walked in a most peculiar manner, and had endless stomach problems. Her companion was a short-haired black-and-white female, for whom we found a wonderful home with Michael's assistant, Rebecca. Hooligan, once alone, made the trip to the vet for her blood tests and shots, lived for years in the laundry room, where floorboard heating was installed for her benefit in the winter, and a flap door so she could come and go as she pleased. She lived on a special diet—expensive, naturally—and eventually her digestive system settled down. She loved to be groomed, which was good, as she was always a mess, drooled continually when you made a fuss of her, and seldom strayed far from home. Then one summer, about half a dozen years after she first came here, she came to the front door of the house, and I let her in. 'Just to cool her jets,' I told Michael—it was a brutally hot and humid summer. She walked around, upstairs and downstairs, tried one of the litter boxes, drank a little water, decided that she liked air-conditioning, and settled onto one of the good dining-room chairs. She has never left since, except for an occasional stroll around the garden in good weather."

Hooligan fit in surprisingly well, partly because she looked exactly like a black fur pillow when she was curled up on a piece of furniture, partly because she wasn't aggressive as cats go, unless directly attacked, partly because given her coloring, she was practically invisible except in broad daylight. With her waddling gait and her curious hair—the spiky tufts of hair coming out of her ears were longer than her whiskers—she was certainly no beauty, but she made up for that in character.

Hooligan.

The next arrivals were very different indeed, more like the arrival of a circus in a small town. With some, there had been a long period of getting acquainted on both sides before they entered the house, a sense that we were choosing the cat, as well as the cat choosing us, but this time it was clearly the cat that made the choice. In England, at the beginning of the Second World War, a vast scheme was put into effect in which whole families were unceremoniously "evacuated" from the big-city slums which were prime, if accidental, targets for German bombing, and dumped on rural households by the authorities, with unpredictable results. We were about to have something like the equivalent of this experience.

It began with a reconnaissance mission. As Margaret remembers, "Mumsie simply turned up at the back door one July morning. I heard her crying before I opened the door, a plate of food in one hand. She was pure white, lean, but healthy-looking. She ate and walked away.

"Maybe she'll be back, I thought, and told Leslie, the same young woman who had found Hooligan with me, to keep an eye out for her.

"Later that day I came home from Rhinebeck and Leslie said, 'I saw the white cat all right, but I think you'd better go round to the front of the house and take a look, that's where she is, with her two kittens.'

"She had two tiny, pure white kittens, one quite bold, the other very shy. For a couple of days they all lived under the porch, but then the remnants of Hurricane Bertha came through, with heavy rains. I had gone off to an event for three days, so I called Michael and asked him to get the three of them onto the enclosed front porch. 'Are you crazy?' he asked. 'We have all that fabulous porch furniture out there, it'll be ruined. Besides, we don't know anything about these cats, they could have all sorts of diseases. Anyway, they're wild, I'll never catch them.'

"'Get Trina,' I said. 'She is helping in the barn this weekend, and she'll give you a hand. And make sure they have food, water, and a litter box. Oh, and throw some sheets over the porch furniture. Gotta go, have to walk the cross-country course.'

"I called later. 'They were all washed away,' Michael said. 'We couldn't find them. We tried, Trina and I spent hours.' There was a long silence. 'Okay, okay, we got them in eventually, just kidding.' The kittens had apparently vanished down a drainpipe in the pouring rain, while Mumsie, not one to panic, pretended not to notice their absence.

"Mumsie immediately realized her good fortune. She had found foster parents for her babies, freeing her up to do all those fun things, like hunting, lazing around in the sun, grooming herself, playtime—after all, she was young, it was her first litter. You have to hand it to her, she was a good mother—she had found her kittens a home—but she wasn't all that interested in feeding them anymore, since there was someone else to do that for her now. A couple of months went by, Mumsie had her vet visit, friends came by to look at the kittens, and the nights started to get chilly. 'We're going to have to bring them indoors off the porch,' I told Michael. 'First of all, it's getting cold, and second, the kittens have to get used to living inside, since they're both going to be indoor cats.'

"Michael blanched. 'What, are you crazy?' he asked. 'Are we going to have three more cats in the house? The entire place will be covered in white cat hair! Oh my God!'

"But they came in, the kittens tore around nonstop all the time, they amused us for hours, and they did indeed cover everything in the house with white hair. Before Thanksgiving, the two of them went off to new homes, one to a colleague of Michael's at Simon and Schuster, the other to our doctor in Poughkeepsie. Mumsie cried for one night, then seemed to forget all about them and settled into the good life. She quickly became a favorite of Michael's, who, I have noticed, falls into a pattern: loves the females in the end, and never really warms up to the males. Mumsie loved to go out all day hunting, and didn't like new arrivals on her turf. She won some, lost some, and staked out the dirty clothes basket to sleep in at night."

Mumsie's arrival inevitably changed the feline pecking order, as any new cat will, but this was aggravated in her case by a certain noticeable sense of entitlement on her part. Perhaps it was that as the mother of two rambunctious twins (for a while we referred to them as "the Terrible Twins"), she regarded herself as having a special status that set her apart from other cats. Certainly she had not arrived looking as if she had spent a long time in the woods and fields foraging for food—it seemed more likely that, as is so often the case, the birth of her kittens was more than her owner had bargained for, and she was simply thrown out of the house, or more likely taken a good way from the house and turned loose, kittens and all.

It's hard to know what goes on in people's minds, but to a lot of people kittens are the straw that breaks the camel's back—they're simply not prepared to deal with them, even though the simple solution is to have a female spayed, but then a lot of people are unwilling to spring for that too. . . .

In any case, one way or the other, Mumsie had been ejected from her home with her kittens, and promptly found a new one, so she did not arrive laden with a whole lot of physical problems and diseases. It must be said that she groomed both herself and the kittens to gleaming white perfection before making an appearance at our door—neither she, nor they, presented the usual bedraggled picture of strays, however hungry they may have been. Mumsie would, in any case, soon prove that she had considerable talents as a hunter.

Neither Chutney nor Hooligan seemed to have much desire to kill things—"Live and let live," might have been their motto—but Mumsie was a lethal weapon when it came to mice, moles, small snakes, and, unhappily, careless songbirds.

Though she was poorly camouflaged—except on snowy days, you could see her from a mile away—she regarded the birdbath in front of the kitchen windows as her particular hunting area, and was quite capable of leaping up and catching a bird just as it landed or took off from the rim. She invariably brought small, brightly feathered corpses, or odd bits and pieces of rodents back to place on the door mat as a present for us, a kind of "thank you" gift for taking her and her kittens in during the storm.

One would have thought that the sight of a pure white cat sitting next to a birdbath and staring up at the sky intently might have warned any bird off from the idea of a bath, but perhaps birds are color blind. Given the fact that Mumsie was snow white, relieved only by a pink nose and pink tips to her ears, hiding in the greenery didn't do much to conceal her, and she mostly didn't bother. She just sat there patiently, until something feathered or of the rodent family came within range, then swooped on it with truly astonishing speed. It was easy to see how she had managed to keep her kittens fed out there in the wild—as hunters go, she was a pro, though now that she was fed two squares a day, she clearly did it *pour le sport,* or perhaps to keep her hand in, just in case.

When she was kept indoors for some reason—usually because we had taken pity on the birds—she sometimes sat on the kitchen counter, next to the sink, staring out the window longingly at the birdbath, eyes focused on each little feathered bather with sheer hatred in them, tail flickering with impatience. In every other respect, she was the nicest of cats, even though she was a menace if you happened to be wearing anything dark. We invested in small vacuum cleaners—the best was the Royal brand—and boxes of rollers with gummy paper for picking up lint and hair, the most effective ones being those sold at the vet's, not surprisingly, but still, every time you put on a dark blue sweater, or a blue blazer, or black trousers, you saw evidence of Mumsie's presence, and being a cat, there was naturally nothing she liked better than sneaking into a closet when the door was open and bedding down for a nice, snug nap on your darkest sweaters. No matter how vigorously Mumsie was groomed and brushed, she had only to rub against your ankles, if you were wearing black socks or stockings, to leave a whole layer of white hair on them. Since she enjoyed human contact, there was always plenty of vacuuming and pet-linting to be done in her wake, to which Mumsie was totally indifferent—like most attractive females, she was always perfectly certain that she was welcome everywhere. People with allergies to cats she singled out as her particular favorites, needless to say.

All by herself Mumsie disproved the adage that "all

cats look gray in the dark," an old English saying (a favorite of Margaret's father) the gist of which is that it doesn't matter all that much whether a woman is attractive or not, since you won't be able to tell the difference at night. Mumsie, however, did not look "gray in the dark" at all, she glowed a spectral white. It was possible to trip on Hooligan in the dark, and we not infrequently did, but Mumsie was as visible on even the darkest of winter nights as the cat's eyes on the English roads, and impossible to trip over. She did, however, blend in with the bedclothes, so it was hard to tell the difference between Mumsie and the sheets, and, as it happened, sleeping on—or preferably, in—the bed, or between the pillows, was one of Mumsie's favorite occupations. Perhaps it was the loss of her kittens, but Mumsie was exceptionally needy when it came to companionship, and quite the reverse of a standoffish cat. With or without even the slightest encouragement, she was on your lap, or on your chest if you were lying down, or on the pillow with her nose in your ear, breathing away gently, if you were asleep. To visitors who weren't used to cats, this was not necessarily the high point of their weekend.

Her relationship with Chutney was friendly, without territorial fights or jostling for position. Mumsie was apparently willing to concede his special place in Margaret's heart, provided her right to a position on certain chairs, and to a place on the bed, was not interfered with. A good long snooze in the laundry basket was her idea of heaven.

Although the picture of the cat with a dish of cream is firmly fixed in people's minds, Mumsie was in fact our only cat who not only liked milk, and of course cream when we had it, but expected, rather imperiously, to be given a saucer of milk every time we had tea or coffee. Failure to produce a dish for her when you took a carton of milk of out of the refrigerator brought on a storm of angry protest and complaint, and Mumsie was nothing if not vocal.

Breakfast was, in fact, her favorite meal of the day. Apart from milk, she expected to get a dollop of butter or margarine, and when possible a few pieces of bagel or—treat of treats—a raisin-bran muffin. She liked to eat breakfast seated on the table, between the marmalade jar and the teapot. At other mealtimes, she was mostly content to stay on the floor, but she seemed to think of breakfast as a family meal, and expected her place in it at, or rather on, the table. After breakfast, whatever the weather, she wanted to be let out for a leisurely patrol of the grounds, just in case there were any songbirds or rodents on the loose at low altitudes. Sometimes she would stay out until quite late in the evening—we have a fair number of acres to patrol—but you never had to worry about locking her out at night, since she quickly mastered a way up to the roof of the house, which involved climbing a tree and jumping onto the kitchen roof, then making her way up one side of the roof to a dizzying height and down the other, to appear eventually at our bedroom window, tapping her claws against the glass for admittance.

Even when the roof, which was fairly steep, was covered in snow and ice, Mumsie could still manage her way up and down the shingles. She found a way, too, of sheltering herself on a narrow ledge below a small skylight window above one of our doors, a precarious position from which we often struggled to rescue her at first with an extended ladder and a pair of thick leather work gloves the moment the cry went out, "Mumsie's on the roof!," until it became apparent after numerous rescues that she was there of her own free will, and could get down by herself somehow whenever she wanted to.

Occasionally, when she had vanished and an area search had produced no results, even at mealtimes, we would happen to look up while walking down the hallway, and there she would be, at ceiling height, looking in through the decorative window above the front door. To say that Mumsie had a head for heights would be putting it mildly—she could climb trees to a remarkable height, and go far out on the limbs, and as she grew older and more cranky would frequently pursue interloping cats far up into the big hemlock beside the barn, then plant herself in a convenient, strategically chosen branch to prevent their coming down. At such times, she ignored meals, weather, and pleading from the ground, determined to protect her turf, or perhaps, as she saw it, ours.

It was, in fact, easy enough to see that Mumsie *did* feel a sense of responsibility. Feckless as we were, in her view, we were likely to let the place be overrun by intruders, but fair weather or foul, Mumsie patrolled it for us, even walking the fence lines, tiptoeing along the top rails, in search of mice, birds, baby rabbits, and of course other cats that might take advantage of us. She liked to start out at the birdbath—scene of so many bloodbaths—then make her way around the flower garden, then go around the barn—rather more cautiously here, because there were often other cats in the barn, auditioning for a place in the tack room, or the laundry room, or ultimately, of course, the house—then from there out to the paddocks to inspect each one, and finally back again to the house, visibly satisfied with herself at a job well done.

Fatigue, bad weather, the occasional cold—Mumsie was subject to sniffles and sneezing—were never allowed to interfere with her self-imposed duty. She never went farther than the indoor riding ring, beyond which there was second growth, dense bushes, miles of riding trails, and the unseen presence of more heavyweight threats, like packs of stray dogs and coyotes. Mumsie's territory was large, but always in sight of the house, in case she had to beat a quick retreat. She had survived in the wild with her kittens, and had no desire to revisit it.

In fact, the only animals before which she retreated were skunks—for obvious reasons—and raccoons, which are more than a match for a cat. Possums she ignored, as if they were beneath her dignity, and wild turkeys she watched with interest, but since they traveled in a flock, dominated by several large male birds, she could tell that the chances of her bagging one to bring home were slim. You could see that she was *tempted*—birds were birds, after all—but despite her taking up her attack position, belly flat on the ground, tail swishing impatiently, chattering her teeth, the turkeys were not impressed or frightened, and she never charged head-on into them, wisely, in our opinion. Small snakes she was quite good at dispatching; larger ones, she was wise enough to leave alone. In any case, every day was full of new adventures for Mumsie, even if they only consisted of teaching new cats on the property who was boss.

The ability to switch between the two modes of a cat's existence was remarkable. Stretched out on your lap before the fireplace, or on the bed, Mumsie seemed like the ideal housecat, as tame and relaxed as could be, but the moment she stepped outside the house, she reverted to her wild hunting state, climbing up and down roofs and trees and defending her turf tooth and claw. Most cats become—or remain—one thing or the other, but Mumsie was comfortable in both roles.

One thing she was particularly good at was calculating the odds. So long as potential rivals stayed outside or in the barn, Mumsie treated them like the enemy, and harried them, whenever possible, into the uppermost branches of the trees, sometimes biting them quite badly, and quite often taking a bad wound herself. But once one of the outdoor cats was allowed indoors, Mumsie grudgingly, gradually accepted the newcomer, and made the necessary adjustments to the all-important social order. She was not entirely reliable, and a certain amount of infighting went on, but it was seldom the open warfare with which she defended her position outside the house.

Fierce as she could be, she adapted to the inevitable— something we could all learn from.

Anyway, the next "inevitables" were bigger than she was.

6. "The Stonegate Strays"

Jake

Over the years, they mount up, the strays who have arrived and stayed, the ones who have arrived and moved on elsewhere, the ones who never quite worked up the courage to appear at the door. After Chutney, Jake was Margaret's favorite. There was no courtship with Jake, no hanging about in the weeds, or the woods, or under one of the outbuildings waiting to be let in. In Margaret's words, "He was at the front door with a companion one morning. He was that lovely gray, like a Russian blue, his friend was a hodgepodge of shades; they were gaunt and very hungry. He was game to come indoors immediately, with his snaggle tooth and little goatee. I named him at once. I felt this might be pushing my luck, so I put them both in the garage. A good move, as when I picked them up, I discovered they were full of ticks. A part-time woman who was working in the barn said she would love to have the female, and took her home that day, but three weeks later brought her back, saying she was going to have kittens. 'Lucky you,' I said. 'They'll be loads of fun.' 'I don't want her,' she said. So I took her down to Mike Murphy. It was a long time ago, and his small animal practice was just starting, and the waiting room walls weren't covered with photos of cats and kittens looking to be adopted, as they are now. 'She's a nice cat, I can find her a home,' he said. 'Just don't tell anyone, or I'll be swamped with strays!' Jake did not pass his vet exam with flying colors, but I kept him anyway, telling Michael, 'He's terminal, he'll only be here a short time.' Over the many years that Jake was with us, Michael repeated over and over again the 'he's terminal' story, pointing out each time that Jake was still with us. 'And,' Michael would add, 'he still has scurf!'

"Jake was very nurturing of anyone, and he would appear from anywhere in the house at the first hint of a sigh, a moan, or a tear, butting his big head against you, walking along one side and down the other if you were lying down, purring loudly and frequently touching your arm with a paw. When Michael arrived home from cancer surgery, Jake was in his element. 'Get him away from me,' Michael whispered. 'He hates me, he always has, he's going to sit on my head and smother me, get him away!'

"Jake was very attached to me. Possessive, sleeping pressed against my side, or, when Michael was away, against my back. He would walk beside me outside, like a dog. He grew extremely fat, to the point where he could get up into the hay loft, but not down, so we learned to keep a fence board handy to act as a runway for him on the descent. Michael says you can always tell 'a Korda cat'—they have small heads and huge bodies. Jake's body let him down in the end. I cried for days.

"Mr. McT"

"Mr. McTavish (also known as 'Mr. McT,' 'Mr. McTiggy,' or 'The Thug') appeared in our driveway one winter morning. Maybe someone had dumped him from their car or truck, or like Hooligan his owner had moved away and left him behind here, or some place nearby. He was a black and white short-haired cat, with big yellow eyes, definitely an adult male. He would come and go—weeks went by sometimes without my seeing him. I fed him whenever he was here. But he grew thinner and thinner. Fleas, worms, and ticks, no doubt. His eyes were runny, his coat dull. Then one June evening he appeared and I let him into the kitchen and fed him a large meal. Michael—and the other cats—were bug-eyed. 'You can't do this,' Michael said. 'It's madness. You can't expose our cats to him, and what about me? He could have—'

"'Everything,' I said, 'but it's only for a minute. I'm going to put him in the garage overnight, and take him down to Mike Murphy in the morning.'

"For the first three weeks, he lived behind the washing machine and dryer, he ate little, and was very smelly. Our housekeeper at the time did not like him, but she didn't stay long. Finally, he came out from his hiding hole behind the appliances, started to eat, his coat bloomed and he grew in every direction, but he did not make friends, especially with Jake and Mumsie, who was frightened of him. 'He's a thug,' Michael said. He never wanted to go outside until very late at night, rather like a vampire, appearing on the front porch roof and staring in the bedroom window. 'He can spend the night out,'

Michael would say. 'It'll be good for him, freshen up his coat, he'll get scurfy like Jake if he spends all his life inside.' Needless to say, he never spent a night out. Over the years, he eased himself into Jake's spot in the order of cats, and as Jake's health failed, McTavish became dominant. Years later, love appeared in McTavish's life, but that's another story."

Once they put on some weight, both Jake and McTavish were bigger than Mumsie, but, as is so often the case with neutered male cats, they were on the whole more peaceful and inclined to let the world go by without fussing. Jake was affectionate, at any rate to Margaret, and had the size and the heft to push his way close to her, but he was not inclined to fight, nor, given his size, was it often necessary for him to. He looked, in fact, more threatening than he was, a Mike Tyson with a sweet nature.

For a good many years, until the arrival of Mr. McT, Jake was the only male cat in the house, and rather like a male lion, he expected to be looked after and deferred to by the ladies, and on the whole was. Bulky and dignified, he always chose the most comfortable chair, the plumpest cushion, the place closest to Margaret, for his nap, and expected not to be bothered. His routine included a generous portion of naps. A good night's sleep, followed by breakfast, followed by a good long nap in a warm place, was his idea of time well spent, and although he had the feline equivalent of AIDs, he had a hearty appetite and put on a lot of weight—enough so that the only proper word to describe him was "portly." His small goatee, his long whiskers, and his expression, all combined to give him a certain malevolent look, which was, in fact, deceptive—push come to shove, despite his impressive size, he was chiefly interested in his own comfort, and his relationship with Margaret. Everything else seemed to him beneath his dignity. He did not sweat the details or let small things bother him unduly. At night you could hear him wheezing and breathing loudly on the bed, and the moment he decided Michael was asleep he would move inexorably up toward the pillows, trying to get between us and push as close to Margaret's face as he could, but beyond a certain disdain for careful grooming he had no bad habits.

Of course, one couldn't help noticing that we had gone from being a one-cat couple to a four-cat couple (counting only those cats who lived in the house), and with no end in sight, for the woods and fields were still alive with cats, and hardly a week went by without some newcomer turning up on the porch at night, waiting to be fed. We had not yet reached the stage of our cat-loving neighbors the Lynns, who bought kitty litter by the hundredweight and cat food in bulk, but four is still a lot of cats, particularly when they all get together in one room, or when it's time to feed them. In the morning, when one woke up, it was to see pairs of pointed, triangular ears in every direction, as the cats woke and thought of breakfast, and cleaning the litter trays turned from being a minor chore to a major one.

Nor was feeding them as easy as one might have supposed, since each of them, once they were no longer actually starving, developed a strong preference for one brand or kind of food only—though, like all cats, they were fickle, and often changed their mind on the subject just after you'd bought a dozen cans of what they'd liked before.

One of the odd things about cats—and another of the ways in which they strongly resemble human beings—is that cats will eat anything they can find or kill to support life when they're on their own in the wild, from frogs and snakes to small birds and rodents, only to become finicky gourmets the moment they're indoors and being fed two squares a day from the A&P. It's as if they went from a state of starvation to being a fussy client in a restaurant or a hotel overnight. You might suppose that they would feel grateful for what they were offered, like the children in the parish poorhouse in *Oliver Twist,* but that is to underrate a cat's power of recovery, as well as its strong individuality. Yes, when they're still starving (and eager to please) you can put any old kind of food in a dish and leave it on the porch and they'll wolf it down, including leftovers, but the moment they're inside the house, they will turn their nose up at what they don't like, and walk away from their plate rather than touch it.

Go figure! Why does a cat which has been perfectly content to eat, say, Fancy Feast Flaked Trout for weeks—won't touch anything else, in fact—suddenly decide it won't touch flaked trout, or even *look* at it? The fickle appetite of cats has made the manufacturers of cat food rich, and doubtless always will. As every cat owner knows, saying, "Oh, he'll eat it when he gets good and hungry," is a delusion. Once a cat turns its nose up at something, that's *it*. It can (and will) sit there until hell freezes over, or until you give in (which is likely to come sooner, given a cat's ability to provoke guilty feelings in its owners).

Of course none of our cats is exactly starving, even at mealtimes, since we have three bowls of dried cat food in the house, always kept filled, just so the cats never go short of a between-meals snack. It goes without saying that each of the dried foods that goes into the mixture is especially chosen because it's good for cleaning their

teeth, or rich in vitamins and minerals, or specially for-
mulated for aging cats, or for preventing problems of
the urinary tract, though the cats don't know this, since
they can't read the labels on the bags. Some of the fussier
ones go to the trouble of picking the kind they like out
with a paw, and flicking the rest onto the floor, but even
if they don't stop for a mouthful, the fact that there is
food available at all times seems to have a relaxing effect
on them, though it doesn't prevent them from scream-
ing like banshees when it's time for their dinner.

Jake, it has to be said, was neither a fussy nor a discriminating eater. He ate slowly—perhaps as a result of
tooth problems—seriously and solidly, whatever was
presented to him, bringing to the act of cleaning off his
plate the solemnity of a prayer meeting. Some cats are
easily distracted when they're eating; not Jake—when it
was time to eat, he *ate*. After eating, he liked a healthy
nap, for better digestion. Unsurprisingly, he grew huge.
His tail seemed small for such a large body, and he
walked with the kind of slow, dignified grace that certain
fat men used to cultivate, in the days when no guilt or
shame attached to being fat. He bore, in fact, a certain
resemblance to Orson Welles in his later years, even to
the expression, which was mildly suspicious, and slightly
devilish.

Though Jake was "terminal," dying was not on his agenda, so it came as a surprise when he began to fail. Various treatments were tried, but the effects never lasted long, and in the end it seemed cruel to subject him to visits to the vet and injections for no real purpose. Not that Jake wasn't stoic, in his own way. One evening, he had turned up at the back door with blood dripping out of his mouth, a sad sight, and had to be rushed to the twenty-four-hour-a-day animal emergency clinic, wrapped in a towel, silently drooling blood. It turned out that he had broken a tooth somehow, but he faced dental surgery with a good deal more calm than his owners, and afterward seemed none the worse for wear, though the emergency vet expressed amazement that he had been eating roast beef! Some cats hate a visit to the vet, and fight back tooth and nail, but Jake wasn't one of them. He tended, on the contrary, to go limp, like a protestor in the hands of the police; on the other hand, he clearly didn't like the experience, and doubtless would have liked it less had he known how hopeless his case was.

Jake's end was eventually the result of a kind of coup. He had never really fought for the position of Top Cat, it was more his sheer size and Margaret's affection for him that kept him at the top of the pyramid. The ladies were not a threat to him, though he took good care not to provoke them, and Mr. McT, while he suffered from a certain jealousy, and glared at Jake from time to time through narrowed, yellow eyes, did not actually challenge the only other male in the house. But as Jake's strength ebbed, and he grew thinner and more listless, Mr. McT became bolder, and more eager to take his place. Mr. McT began to jostle and push Jake, until finally, he managed to attack the big gray cat and mutilate one of his paws. Dripping blood, Jake had to be taken back to the vet, but the defeat and the injury at the hands of Mr. McT seemed to strip Jake of his will to live, and, in the end, as his immune system collapsed, he simply gave up.

Margaret adds: "Cat bites and scratches can be serious for people too. I remember early one morning when I was having my breakfast, Queenie jumped up on the table and when I picked her up to put her down, she bit my finger. I didn't pay too much attention to it apart from washing it clean and putting on a Band-Aid, but by late afternoon, it was throbbing and very swollen, so I drove myself into Poughkeepsie to Vassar Brothers Hospital, where of course I waited for ages in the ER, a reason one is often making for not going when one should. Eventually the necessary forms were filled out and I found myself in one of those little curtained cubicles. A very young intern came in with my form in his hand and said that he had noticed that I was allergic to penicillin. 'How allergic?' he asked. 'Try death,' I said. He sighed, 'What a pity, it would have been the best treatment.'"

Mr. McT slipped effortlessly into Jake's place—no doubt where he had always wanted to be—put on weight, bulking himself up almost to poor old Jake's size, and replaced Jake on Margaret's side of the bed.

"The king is dead. Long live the king!" With cats, as with human society, the business of life and succession go on in an orderly way. Whether cats feel grief or not is hard, of course, to say. Certainly they miss a friend when he or she is gone, though by human standards they recover swiftly from grief, but then again Jake's friendship had been reserved for Margaret, rather than his fellow cats.

In any case, it was Mr. McT's moment, and he made the most of it.

7. Mr. McT in Love

Mr. McT had a streak of the bully in him, and
although both the "ladies" of the house were more
than capable of looking after themselves in a scrap, neither
one of them thought herself a match for Mr. McT, and being
cats, it never occurred to them to team up against him—
teamwork just isn't a big cat concept. Hooligan wasn't
afraid of him, or any other cat—she just stood her ground,
howled, and puffed herself up into a huge spiky, inky bundle of fur with which it seemed prudent not to tangle—
but Mumsie was, and made the mistake of showing it.

Not that Mr. McT was looking for trouble—he simply

wanted to have his own way, without any interference—best place on the bed (nearest Margaret), first plate of food at dinnertime, a place on Margaret's lap and scraps off her plate at lunchtime, that sort of thing. Nothing about Hooligan or Mumsie suggested that they saw themselves as Mr. McT's harem, even had he been in a position to need a harem, and on the whole they did their best to steer clear of him, Mumsie particularly.

From Mr. McT's point of view, it was a good life, but for cats, as for people, things change. There was about to be a whole new set of arrivals, all females, beginning with a singularly self-possessed stray who was not perhaps as stray as she first seemed.

As Margaret recalls, unlike that of most strays, who hang about trying to make up their mind about whether to approach closer or not, her arrival was not quite as accidental as it looked. "Kit Kat I know was planted on us by my two barn workers, who lived next door. She had most likely made a nuisance of herself at their house, and they quickly figured out that I would be a soft touch as usual. So surprise, surprise, when I came out to the barn one morning and Leroy said, 'Look, a new kitty, she certainly came to the right place. What do you think Michael will say?' I said, 'You know perfectly well what Michael will say, and isn't that the cat Juan said had come down his chimney last week, and that you said had been hanging around your garden?'

"Leroy started to back away from me, taking his cap off and putting it back on again, and staring into the middle distance, a sure sign that something was not on the up and up. Juan appeared, saying. 'She the kitty who come down chimney, she run around all over the house, we like, but she no stay.'

"Leroy gave him a look and walked away. 'Goddamnit, Leroy,' I said, 'we don't need another cat, you know that.' He kept walking. Kit Kat and I looked at each other. She was very pretty, very feminine, no doubt about that. Orange, tabby, and white. I walked back to the house, and she followed. She came straight inside.

"For a cat, she had a hard time judging distances, and was always taking off from one spot but never making a secure landing where I imagine she had planned to.

Kit Kat, the flying cat

Show off!

China, books, photos, precious small objets d'art all hurtled through the air, and flowers flew out of their vases and lay in pools of water on the floors. 'She needs some sort of tranquilizer,' Michael said, as she continued on her path, stacks of videos tumbling from shelves, pots and jars and bottles on bathroom shelves crashing to the ground. He liked her despite all that, and even overlooked the fact that one day we discovered that she had neatly bitten around the bottom edges of every lampshade in the house. 'Those were very expensive, and we brought them back from Santa Fe,' I said.

"'She didn't mean any harm, she's the pretty one,' he answered, and started carrying her around, and letting her sit on his lap. But she was tough, and swatted at him all the time, often scratching him, and he began to appear with iodine swabbed over his fingers and up and down his arms. She never purred, wore an irritable expression, fought all newcomers, and the vets were always pleased to see her leave their examination rooms. Over time she has slowed down, and we have less damage indoors, but I am always pleased to see her waiting to be let out in the morning—it gives us and the other cats a break."

Kit Kat was certainly pretty, and full of strange and endearing habits—she disdained water bowls, and preferred to drink from a faucet, for example. She tended to lurk on the kitchen countertop next to the cold-water faucet, and would tap your hand when she wanted the water turned on for her. Unfortunately, she did most of her tapping with her claws out, so if you weren't fast enough to suit her, it was back upstairs for the iodine bottle and a Band-Aid. Eventually, it proved necessary to keep another bottle of iodine and a big box of Band-Aids in the kitchen cupboard. Once, she managed to take off Michael's glasses with a single swipe of her paw, sending them flying across the kitchen, but mostly she satisfied herself by leaving a series of crisscross cuts on your wrist that bled for hours and made you look like a failed suicide attempt. She also had a habit of hiding in out-of-the-way nooks and crannies to give you a quick scratch as you walked by. She had enough charm to make these seem liked lovable habits, except at the actual moment when she had her claws in your skin.

Mr. McT gave her a reasonably wide berth—it did not escape his attention that Kit Kat was not a cat to be tangled with. Her expression, even in repose, was menacing. Not surprisingly, she quickly took on the telltale look of "a Korda cat," which is to say a big body and a little head. One of the objects of her persecution was Tizz Whizz, a stray who, after much drama and enticement, spread over a long period of time, had finally taken up residence in the tack room.

Margaret has been grooming Tizz Whizz (literally) for greater things, and calls her "a success story, over almost a two-year period."

"She appeared to be living belowground in the driveway culvert, and continued to live there for some time, causing us a lot of anxiety when after one heavy snowstorm her exit hole was piled high with plowed and frozen snow, and we could hear her cries from deep below. She is probably Kit Kat's mother or sister, as they are almost identical. I ran out through the first winter and left food by the opening. Then she started appearing in the barn. You could not get near her, and she hissed and spat at all of us. But we put a nice box in the hay stall, with a bowl of dry food and water and that was okay until the spring got warmer. She grew thinner. One evening I was sitting out on the mounting block and she joined me, rubbing against my back. I put my hand out to stroke her, and off she flew. So I got a small hairbrush and held it out to her. A long time passed and then she started rubbing her face against it. The brush was the turning point. Gradually, she let me touch her, and even pick her up, but to this day nobody else can touch her. After nineteen months she decided to come into the heated tack room through the cat door we installed for her. She now lives there most of the time, sitting in the sun-filled window, not bothered by the comings and goings of all of us. One day, while we were having our Dunkin Donut break in the tack room, Michael said, 'She looks as if she's ready to come into the house, but then one will have to come out and live in the barn—McTavish gets my vote.'"

Poor Mr. McT! Needless to say, suggestions that he should be turned out of the house fell upon deaf ears, but he was now living with three strong females, none of whom was about to put up with any nonsense from him. He had managed to replace poor Jake, but was still obliged to walk carefully, despite his size and thuggish expression, except when looking at Margaret. His life, however, was about to change radically, in the form of an unexpected May/September romance, proving, if nothing else, that with cats as with people, nobody is immune from love.

Margaret: "Ruby 9/11 is the only cat who has come into our lives without baggage. She was found in the weeds around the pond next to our annex barn, where we store jumps, hay, and equipment, near our neighbors Bill and Lisa, who now live in the house where I found Hooligan abandoned many years ago. They called to ask if we were missing a cat, and I said no. The next morning, I rode by and Bill picked her up out of the long grass. 'Pretty little thing, isn't she?' he asked. She had a pinched little face, huge ears, long, gangly legs. Mostly black, with some pale orange and white. She was so thin, so small. Very young, and I wondered if she was on her own, or if there were more. 'If you could put a can of food and some water in the barn,' I told Bill, 'I'll be back in a while.' Later, I took up some of the necessities of life for a cat—a nice basket, with a towel inside, a food dish and a water bowl, a flea collar, a couple of cat toys, and a brush. She was gone.

"Just as well, I thought, hard to explain to Michael, even though she could have lived up here and been an annex barn cat. I was about to get back into my car, when she walked out from some weeds and started to climb up my pants leg. I picked her off, and realized I was holding on to skin and bones—she weighed nothing. I carried her into the barn and put her down. She wasn't the width of my wrist. She climbed up my leg again, and this time she made it all the way to my neck, purring loudly.

When I got home I made an appointment to take her to the vet the following day. 'I found a kitten today,' I told Michael at dinner. 'She's going to live in the annex.' 'Really? And for how long?' 'That will be her home,' I said.

"The vet visit went well, and she did not even have fleas. That was on a Thursday afternoon. Each day we rode by, this tiny thing would be sitting in the doorway, and on Sunday morning there was a terrible storm. This isn't going to work, I thought, so I drove up, collected her and her belongings, and brought her home. 'Her name is Ruby 9/11,' I said, 'because that's the first date I saw her.' 'But we have four cats in the house already, plus two outside,' Michael said. 'It isn't going to work. She's a kitten, the other cats will kill her.' Michael was putting iodine on his hand, after several scratches from Kit Kat.

"As it turned out, Hooligan and Mumsie were briefly curious, Kit Kat confrontational as usual, but Ruby was smart and subservient, and gave her a wide berth. McTavish, however, fell in love. How could he have been this lucky? he must have thought. They played together, they groomed each other, they ate out of the same bowl, they curled up and slept together, wrapped in each other's front legs. It was a May/September romance all right. 'No more cats, absolutely no more,' Michael cried. 'Have you seen what they've done to the wallpaper in the bedroom? And to the dining-room chairs? And to the green velvet armchair? And to the carpets? I mean, Thom von Buelow would have a fit if he saw this house now!'"

Ruby grew. Unlike the rest of the cats, she didn't put on a lot of weight, perhaps because she was active as only a kitten can be, waking us up every morning with the crash of objects falling to the floor as she leaped and bounced from one piece of furniture to another, without, apparently, looking where she was going. Her specialty was the high jump, leaving a trail of wreckage in her path. She also liked to start the morning at, or slightly before, dawn, by playing with her toys, scattering them noisily across the polished pine floorboards, and hiding them in inaccessible places. Sleep was impossible once light began to come through the windows, however faintly.

On the other hand, in her calmer moments, she liked to cradle herself against the astonished Mr. McT, with her paws around his neck, grooming him gently. In her hands, Mr. McT, with his thuggish ways, who had won his place at the top of the heap by tormenting poor old Jake in his last days, became a new and more gentle cat.

It would be nice to think of his transformation as the perfect ending to a story—love conquers all, even

McTavish—were it not for the fact that Mr. McT was, like Jake, getting older and sicker. He lost his healthy appetite, he lost his impressive bulk and swaggering ways, and finally be began to lose control of his rear legs, so he was obliged to pull himself along with his front legs, dragging his hind legs behind him, and was often unable to get up once he had lain down. He had diabetes. It is possible—Who knows?—that Mumsie, Hooligan, and Kit Kat might have ganged up on him, but Ruby protected him, always staying close by, and looking after him, as faithful as could be. By this time, she was nearly fully grown, a biggish cat, but still very thin and rangy, and she devoted herself to comforting Mr. McT as much as she could, until the final, sad day when the vet who was coming over to put Margaret's beloved horse Missouri down—he was nearly thirty, she had owned him for nearly twenty-five years, and of all her horses he was her favorite, but age and illness had finally caught up with him—agreed to take care of McTavish in the kitchen at the same time, thus sparing him the dreaded trip to the vet, and allowing him to die quietly in his own home, surrounded by the only two beings who had ever loved him, Ruby and Margaret.

He was buried together with Missouri, and the ashes of Chutney and Jake, each in a separate Petrossian bag, at the end of the lawn, where it meets the woods, within sight of the kitchen window, and Margaret was heartbroken, but confident that she had done the right thing.

As for Ruby, although she had lost the one big love of her life (to date), in the manner of cats she did not grieve, or lose her appetite, or sit around looking lost and mournful. To this day, she still rather likes to lie down on the peach-colored chaise longue in the bedroom (on a protective towel, since this was one of Thom von Buelow's pricier fabrics), where she and Mr. McT had liked to cuddle together in his last days, and perhaps she remembers him, or there is still a trace of some familiar scent. . . .

You never know what goes on in the mind of cats, after all.

But in its own way, McTavish's life *did* have a happy ending—he found a good home, and at the end of his life he found love.

Cat or human, that's not such a bad way to go.

8. Tootsie Comes to Stay

Tootsie

There is always another cat waiting in the wings.

Even while poor McTavish was still with us, he would occasionally look out the window of "the library" (which, now that Thom has moved to Tuscany, we have transformed back into "the television room") and see a fleeting white shape on the little porch outside at night.

This was one of our three outside cats, all wild and shy. One of them, Agent Orange, had been around for years. Margaret had ambitions of getting hold of him, but, as she admits, "I have never been able to get close to him, let alone touch him. Every time he disappears for a week or so, I think that's it, he's dead, but he always shows up again. He is a large, bright orange cat, or ginger, as we call the color in England. He looks better during the cold weather, as his coat fluffs up and gives him the appearance of weight gain. Periodically, I put worming medicine in his food, usually in some chicken I've cooked for him. In winter we put doors up on the dining-room porch, with a cat door for him, which he quickly learned to use. Sometimes he'll curl up in a box, which is there for him year-round, or lie on the door mat. But the minute my hand touches the doorknob, he's gone. So I am never able to treat his wounds when he comes back with them, or brush his matted coat, or get a flea collar on him. On evenings I don't see him waiting, I call and bang his plate on the stone floor of the porch. Hoping he will appear before the food freezes in the winter, or is covered with flies and ants in the warm months. And I do wait until I see him before putting out water in the wintertime, since it freezes so quickly. He will run away when I do this, but he comes back."

Agent Orange continues to elude Margaret, and also continues to come back just when you think he's gone for good, usually sitting on the low garden wall next to the herb garden about suppertime. He appears to live under the old lawn shed/garage in which Margaret's midmount

mower and some other equipment is kept, and occasionally Mumsie used to slink away to visit him there.

Tootsie was a later arrival. Margaret described her as,
"Very wild. Pure white, with a calico tail, pale orange
ears, and a dot of color over each eye. She lived in the
cavernous trailer barn, where we set up a warm basket
behind one of the horse trailers, and gave her dry food
every day. I saw her on Michael's office porch some
evenings, where we have another cat basket, and I put
wet food out for her, hoping she would it eat it before too
long. Some evenings I actually saw her, but if she caught
sight of me, or Michael, she was gone. I say 'she,' as white
cats are so often females, but I could have been wrong."

But Margaret was not wrong, as it turned out, for
Tootsie, unlike Agent Orange, finally stepped into the
house with no regrets or fears once the door was opened
to her, shook herself off, took a look around, and
decided to stay.

At first, she was kept in a wing of the house contain-
ing Margaret's "trophy room," hung with ribbons and
silver awards, the "library," Michael's office, a small
kitchenette, and, up a steep, twisty eighteenth-century
flight of stairs, Margaret's office and a bathroom. Once a
separate house about a quarter of a mile away, it was
transported on log rollers and fastened to the bigger

house early in the nineteenth century, presumably to cope with the needs of a growing family, and a single door now connects it to the hallway of the original house, so that it forms a kind of separate, self-contained two-floor suite, the wide-plank, eighteenth-century floors on different levels than the rest of the house, which results in odd, mildly dangerous steps where you least expect them, and very low ceilings. Tootsie's litter tray was placed in the kitchenette off Michael's office, much to his horror, making a total of three in the house (our neighbors the Lynns once had twenty-eight), and her food bowl and water bowl put in the fireplace, on a tray. The general idea was that this was a temporary measure, following which Tootsie would be adopted by some cat-loving friend, or would integrate herself seamlessly into the household. Like so many temporary arrangements involving cats, it gradually became permanent.

Tootsie settled in very quickly, and made the rooms her home, with the result that for some time she seemed unaware of the fact that there were other cats in the house, while they seemed unsure of her presence. Margaret had intended her to stay until she warmed up, but it soon became evident that she had no intention of crossing the threshold into the hallway and entering the rest of the house, while the other cats, once it became clear to them that she was really there, proved to be equally unwilling to cross over into what was now clearly Tootsie's territory.

This produced something of a Mexican standoff. The door to the hallway, which initially we kept closed, was like Beirut's famous Green Line, or the Berlin Wall, or the Israeli fortifications on the West Bank, a barrier so uncrossable that when we finally lost patience with keeping the door closed, neither Tootsie nor the other cats would cross the threshold once it was open. Tootsie had no interest in exploring the rest of the house—indeed she may not have realized at first that it existed—and apparently still less in meeting any of the other cats. As for them, they tried to pretend that nothing was out of the ordinary, and that they had never wanted to go into the library or through Michael's office and up the stairs to Margaret's in the first place, even though the former and the latter had been among their favorite places to sit.

Tootsie's favorite place was in the library/television room, on the back of one of Thom von Buelow's most expensively upholstered armchairs, covered in beautiful white chintz with roses. There she could look out the window, or turn her head and survey the doorway, commanding the high ground in case any of the other cats entered the trophy room from the hall. She showed no desire to go outdoors again, and raised a terrific fuss if the door to the porch on which she had once sheltered was even opened a crack, and apparently did not feel any need to exercise, beyond the occasional wary stroll to the litter tray or the food bowl. What she liked best was when we had dinner and watched television—she was very content to sit on the top of the sofa, behind our heads, or to sit on Margaret's lap. On nights when we did not eat dinner in the television room for one reason or another, she howled and complained noisily, furious at the disruption to her routine and at being deprived of human company. She was, obviously, a "people cat."

Very occasionally, Mumsie, one of the braver and more assertive of Tootsie's sister cats, made a forced entrance into the library, but this usually produced a fight, and since they were both white cats, a blizzard of white fur. Eventually, Tootsie would come out of her domain and sit in the hallway at breakfast time, a foot or two from the door, howling piteously just to make sure we didn't forget to bring her "room service," but always making sure of her line of retreat first. Once or twice, to break the ice, we carried her upstairs, despite wails and moans on her part, and plunked her on the bed, where she would sit quite happily for a few minutes (thinking, no doubt, "This is the life!"), until enough of the other cats appeared to make her feel surrounded, then she would choose a good moment when their attention was directed elsewhere to jump off the bed and scurry downstairs, back to the safety of her own apartment.

It's important to understand that cats are quick to form territorial rights to certain places and positions in a room, to which they attach enormous significance, and of course a newcomer, even with the best will in the world, is only too likely to upset the existing arrangement and cause chaos, rather like a new guest settling into a boardinghouse and sitting down in "somebody else's" chair. In our case, the bedroom, a sketch of which appears on the next page, had been very effectively divided into certain key positions—Hooligan, in her basket beneath the window, Ruby on top of the television cable box, which was formerly Kit Kat's place, until she gave it up to Ruby, Kit Kat in the laundry basket, which used to be Mumsie's place, until she apparently traded it to Kit Kat, and Mumsie herself on the bed, as close to the pillows as possible.

These positions had not been reached without many disagreements in the past, and of course death had played a role in freeing up certain of them—Jake and McTavish had fixed favorite spots too, and had held on to them by seniority, size, and a certain sense of male superiority.

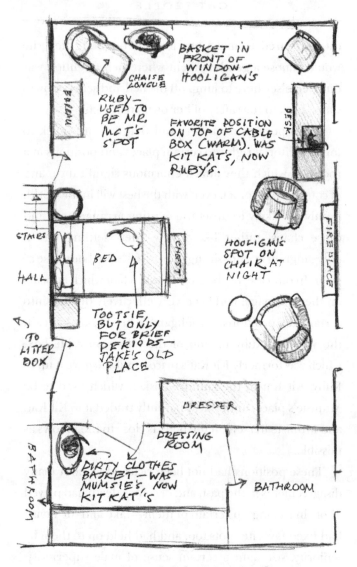

CAT PLAN — MASTER BEDROOM

Particularly choice spots, like the basket below the window were much sought after and envied and only after McTavish's death did Hooligan feel free to make it her own. Until Kit Kat herself gave up the cable box as a place to nap or sleep (very high ground, plus the no doubt pleasing electronic warmth), it would have been unthinkable for Ruby, as a newcomer, to sit there. Also there was no challenge for the place from the other cats, since it required a couple of high jumps to reach—really, Kit Kat and Ruby, big jumpers, were the only contenders.

It is all very well to love Elsa in *Born Free*—one would have to have been "born dead" not to shed a few tears at the ending—and to accept at face value the message that lions are, in fact, or ought to be, "born free," but lions, like all the members of the cat family, big and small, are ruled by precise social obligations, which they take very seriously, and, when necessary, enforce with extreme violence. They are born into a society with rules as rigid as those of feudal Europe, with mutual social obligations that mean a lot to them. Far from being "born free," cats are born into a world of demanding rules, a whole litany of "do's" and "don'ts" that are bred into them, and also, no doubt, passed on from mother to kitten, and reinforced as necessary with a sharp slap. We like to think of animals as "free" in ways that we are not, but one of the benefits of owning cats is the ability to perceive that life is just as complicated for other species as it is for ours, and the need to figure out "the right thing to do" in complicated circumstances, just as worrying as it is for us. Cats do not have the equivalent of an Emily Post to guide them, but etiquette and preservation of social order matters just as much to them as it did to Ms. Post, and they spend a good deal of their time trying to figure out where they belong, and to combine feline good manners with self-interest. A little like us, you might say?

In any case, the objection to Tootsie was not so much her presence upstairs per se, but the question of where she would fit in—a real estate problem, really—while for Tootsie the difficulty was to avoid making a wrong step, or faux pas, which might offend the other cats and spark off a fight over territory in which Tootsie could find herself fighting one against four, with predictable results.

Anybody who has been to boarding school, or college, or served in the armed forces, will instantly recognize the difficulty that a newcomer faces when entering *any* place in which the chairs, the bunks, the bathrooms, or the closets have already been divided up. Tootsie's choice of the bed—well, she didn't actually *choose* it, since we plunked her down on it—upset both Mumsie and Kit Kat, who felt that they had long since earned "bed privileges," and the resulting level of hostility, though silent, was enough to make Tootsie run for cover.

Creatures of habit, cats like to have things in the place to which they are accustomed. In the kitchen, mealtimes have to follow an exact *placement,* rather like a formal dinner with place settings. Kit Kat eats on the countertop, to the right of the sink, next to the toaster oven. Hooligan eats on the floor, in front of the refrigerator. Mumsie would either eat on the left of the countertop, separated from Kit Kat by the sink, or on the kitchen table. Ruby, as a newcomer and a slow eater, gets the floor, as far away from Hooligan, who is a fast eater, as possible, since otherwise Hooligan is apt to come over and push Ruby away from her food so she can steal a second helping. Tootsie, of course, gets room service. Changes are not appreciated, and the order of who gets fed first is equally sacrosanct.

Eventually, Mumsie and Tootsie made peace rather tentatively—they occasionally even shared the television room sofa, formerly off-limits to cats, covering the most expensive piece of furniture we own with white hair, but their eyes always remained firmly fixed on each other, just in case. They sat at opposite ends of the big sofa, facing each other, completely immobile, like two small white Sphinxes, waiting to see who made the first move.

In the meantime, Tootsie became a kind of fixture.

She shows no desire at all to expand her horizons—she has never made it to the dining room or the kitchen, she has her own private litter tray, and lives altogether like some kind of mystery passenger who never leaves her cabin on a luxury steamer or the old Orient Express, receiving her meals on trays, and never emerging to mix with the other passengers. There is, in fact, something of the grande dame about her, with the odd splodges of color above her eyes, and her ability to sleep for about twenty hours a day.

And still they came. Margaret never stops looking for new cats.

"Last fall after a long season of eventing with the horses, my barn manager, Toby, and I decided it would be fun and something quite different to go to the cat show at the Civic Center in Poughkeepsie. I particularly wanted to see some of the more exotic breeds, seldom being exposed to any cats but the ones arriving at the door.

"But after we spent some time walking up and down the aisles, oohing and aahing at this breed or that and shaking our heads at others, we naturally gravitated to the area where the different adoption agencies were located. What a mistake! Especially as I was on the rebound from losing both Mr. McT and Mumsie and felt particularly vulnerable. Each kitten, pairs of kittens, families of cats, old battered ones, young skittish ones, I found myself saying to Toby, 'We've got room for this one or perhaps those two.' She smartly moved on to check out the 'accessories' area and by the time she came back I had two black and white eight-week-old kittens, brother and sister, put to one side, while I continued to look for more. 'Don't think Michael is going to like this much,' she said. 'Okay, let's stop here.'

"I filled out the forms and answered all the questions that went along with the adoption. Telling everybody who was now stopping by to look at the two adorable kittens, tussling with each other, 'They're mine. I've already named them, they are called Bonnie and Clyde.' Everything was going very well until it came to the form of payment for their adoption fees. I immediately whipped out a credit card, being a plastic person, only to hear that cash or a personal check would do. I heard Toby's voice behind me, 'I have cash'—what a relief— into one of the agency's travel boxes they went, perhaps not quite so energetic as they had seemed earlier, and I was asked if I would like to strike the gong. 'We strike the gong each time a cat or kitten is adopted,' I was told, 'so you get to hit it twice!' What a moment as I banged away, heads turned, people clapped and cheered.

"Toby was in a little bit of a hurry to be on her way once we arrived in the driveway and Michael came out of the house. So I was left standing with the cat box. 'Hi, guess what?' 'I've guessed,' Michael said.

"This story did not have the happy ending I had hoped for. Bonnie and Clyde who had grown quieter and quieter on the ride home, and who made visit after visit to the litter tray from the minute they got out of the carrier in the kitchen, through the evening and night, were at our vet's doorstep first thing the following morning. Where it was very quickly discovered they were infested with a type of parasite, hard to get rid of and not a situation I felt I should expose our other cats to, so I had to call the agency who agreed to collect them from my vet's and return them to the foster home where they had been living prior to the cat show."

Thus life goes on, with a new mix of cats, a changed cast of characters. Out of the blue, as in all life, awful things happen. As Margaret says, "We are never prepared for death by accident. If there is an existing illness, upcoming surgery, old age, then we have a period of time to accustom ourselves to what is coming, and likely make arrangements.

"But one Wednesday lunchtime last September, just as Libby and I were sitting down in the kitchen, Juan's wife, Dolores, who takes care of the house, called out my name and Mumsie's from the front door. Lauren, who worked for us in the barn, stood on the porch with Mumsie's body wrapped in her sweater, a stranger standing behind her. 'She ran out across the road, right in front of his truck, there was nothing he could do, it wasn't his fault,' she said. 'Not Mumsie, not Mumsie,' I kept saying. 'What am I going to tell Michael? She was his favorite. Not Mumsie. She never went across the road, not in ten years. Go away, go away from here,' I shouted at the man. 'Go away.'

"I carried her around the house wrapped in the sweater, to all the places she had known over the last ten years, stroked her coarse fur, and put my face against her side. 'She still has the same wonderful smell,' I said to Libby, who was following me around, rubbing my back each time I started crying. 'This hasn't happened, it hasn't happened.'

"But it had. And Toby dug a grave for her next to Missouri and Mr. McT. There is no marker, no headstone. There are none for any of our animals. There is no need.

"I carry them with me everywhere, they are in my heart."

From time to time, Margaret misses Mr. McT too, or Jake, or Queenie, or Chutney, or poor old Irving, who was the first to put his paw in the house, not to speak of those who really just passed through briefly, Bigfoot, for instance, or the unfortunate Mrs. Bumble, and says how much nicer it was when Jake was beside her in bed, or when Chutney would sit on her lap while she drank her tea, but there it is—we have five lady cats, one of them a kind of feline Blanche Dubois figure, all sharing a big house, plus two more outside.

That might seem a lot of cats to some people, but of course to many other cat people it's hardly even a drop in the bucket.

9. *La Chatte Transatlantique*

entre-chat

It helps to have friends who are as crazy on the subject of cats as oneself (or crazier), and this is, very fortunately, easy enough to achieve. Complete strangers, of course, are very often even crazier. How else to explain the fact that a company called Genetic Savings & Clone, in Sausalito, California, will clone your cat for fifty thousand dollars, and is said to have a long waiting list of customers?

Well, of course, no doubt it *does* make a kind of sense in southern California, land of *The Loved One* and Whispering Glades, and of the chimpanzee funeral at the beginning of *Sunset Boulevard*. After all, people who are prepared to pay thousands of dollars for the stylish funeral of a beloved pet are surely getting a better deal by spending money to clone one—if you truly love little Tuffy, why not simply have him or her replicated, if necessary over and over again? Eternal life for pets, more or less, if you can afford it—anybody who has ever read *After Many a Summer Dies the Swan* can only wish that Aldous Huxley were still alive to write about it, but he died, poor man, when cloning was still the stuff of science fiction, particularly since hardly any English man of letters and science, except perhaps for H. G. Wells, would have been more eager to be cloned, given the opportunity.

For those for whom cloning is a little too pricey, there are painters who specialize in doing oil paintings of cats, so you will always have a portrait of your cat to remember it by, as well as sculptors who can replicate your cat in bronze or marble. There is, in fact, a very talented cat portraitist who lives and paints not twenty minutes from us, but so far we have not taken any of our cats to her studio, though Thom von Buelow *did* buy us a life-size bronze cat in which to keep Queenie's ashes, which sits just below the birdfeeder, without, apparently, frightening off the birds, though it's more a generic cat, really, than an actual attempt to reproduce Queenie. It has a bushier tail than she did, and two front legs; still, it was a lovely, if expensive, thought.

We try to keep in communication with friends who have adopted one of the local strays over the years—Michael's retired assistant Rebecca, our friend Dick Olpe, Theresa Horner (who got one of Mumsie's pure white kittens). There's a whole group of people out there with cats they were persuaded to adopt, or simply fell in love with at first sight while visiting us. Then there are friends whose obsession with cats more than equals our own.

For example, there's Leila Livingston, a neighbor of Margaret's from way back when Margaret and Burt first moved into their Central Park West apartment, B. I. ("Before Irving"), or more years ago than any of us would care to count at this point, frankly. Leila was then hardly more than a child, and appeared, waiflike but determined, feet planted permanently, even then, in the third position, to protest the fact that some plumbing problem in Margaret's bathroom had led to a raging flood downstairs. Only in New York City, perhaps, can this kind of encounter lead to a long friendship, but anyway it did.

Leila, as it turned out, would become a ballet dancer, fluent in French, a brilliant cook, and a certified cat lover. When Michael came to meet her many years later, her command of French awed him, as did her fund of literary cat anecdotes, such as the one about Mallarmé's cat,* whose name was Blanche, and who liked to sit in the window of the great French symbolist poet's apartment, looking out at the street. One cold, miserable winter night, a bedraggled, starving alley cat looks up to see Blanche, plump, warm, and fluffy, sitting at the window. The alley cat creeps closer and says, "Comrade cat, how can you live in luxury as a pet when your brothers and sisters are out in the street starving?" Blanche frowns. "Have no fear, comrade," she replies. "I'm only *pretending* to be Mallarmé's cat."

This is the kind of animal story that not only captures a certain feline view of the world—How many of us, after all, suspect that our cats are only pretending to be domesticated pets for so long as it suits them?—but also the typically French bourgeois attitude toward what used to be called "the class struggle," in which a degree of more or less feigned sympathy for the suffering of the lower classes, despite one's own comfort, is supposed hopefully to serve as protection against proletarian reprisals in the event of revolution. Without an understanding of this social mechanism—the need to shout from time to time,

---

*Also quoted in a slightly different form in "Anecdotal Evidence," by Eliot Weinberger, *Conjunctions*, 2003, and attributed by him to André Malraux, war-hero, novelist, art historian, and minister of arts and culture under de Gaulle.

"But I've always been on your side!"—which goes back to the mid-eighteenth century, much of French politics remains baffling and incomprehensible to Anglo-Saxons.

England too proliferates in talking pets with a social message (though not the same one, needless to say): "I am his highness's dog at Kew, pray tell me, sir, whose dog are you?," a clear statement of class-conscious snobbery, was actually engraved on the gold collar of the Prince of Wales's dog (*not* Charles, but George III's eldest son, who went on to become Prince Regent, then George IV).

Then too, though the English think of themselves as great animal lovers (and indeed are generally assumed in the rest of the world to be loony on the subject of cruelty to animals, assaulting Arab donkey owners and Italian drivers of horse-drawn carriages and abusers of animals everywhere), hardly anywhere in England is the French love of pets equaled—or would it, indeed, be allowed. Nobody in France thinks anything of bringing a dog into a restaurant—elderly ladies carry their little dogs into elegant restaurants and cradle them in their laps, feeding them choice tidbits, while bearded gentlemen wearing berets, with napkins tucked firmly under their chins, still sit at café tables with their dogs seated in the opposite chair waiting for a treat. In England—let alone hygiene-conscious America—none of this would be tolerated for a moment. As for cats, every concierge, and sometimes, it seems, every grocery, boucherie, and fish shop in France, has a resident cat, mostly cross, over-fed, and spoiled. Louis XV, when he made his first formal appearance before the court at Versailles as a boy-king, carried his favorite cat in his arms, and was much admired for doing so. The king retained a lifelong fondness for cats, and a good many of them appear in the paintings of his mistresses. In general, the French live on more intimate terms with their pets than the English do, much of the intimacy of course centering, as you would expect, around food, and are astonished and outraged to discover that the English, of all people, consider them cruel to animals.

To say that Leila was sophisticated would be putting it mildly. Indeed she succeeded at the remarkable feat of being witty in two languages (most people find it difficult enough to be witty in one) and of being completely *au fait* with two cultures. Eventually her ballet career came to an untimely end due to an injury (though one look at the way she stands, even at the stove, would reveal her ballet training), and she went off to Paris to attend cooking school, taking her sixteen-year-old cat Cleopatra with her—it never occurred to her for a moment, as a matter of fact, to leave the cat behind.

Cleopatra had been rescued by Leila's mother from

the courtyard of their building, along with a sister, which they named Medea Odille, who went insane, biting and hissing, and proved too resistant to becoming a pet. Cleopatra, on the contrary, adapted quickly to petdom, although she shared the apartment with two parakeets, and liked to sit on their cage, glaring in through the bars and terrifying them into motionless, frozen silence for hours at a time. She also shared the apartment with a French poodle named Antonio, and an orange cat called Mephisto, for many years. Timid, except when it came to parakeets, Cleopatra never ran out into the hall when the door was opened, and vanished at the sight of a stranger, so it could hardly have been expected that she would be a successful traveler. However, when Leila went to Paris, Cleopatra nerved herself up for the voyage, and crossed the Atlantic like a veteran.

Even in those days, before the big security flap, international airlines were ambivalent on the subject of pets, though in principle most carriers allowed one animal per cabin, but Leila simply boarded the plane, shoved Cleopatra's case under the seat in front, and sat the cat on her lap across the Atlantic, having prepared her for the journey with a quarter of a five-milligram Valium.

"On planes she would spend most of the flight asleep on my lap, even though she was meant to be in her kitty bag under the seat. Being a cat, no matter how long the flight, she would never go to the bathroom, even though I always took some kitty litter with me, and an improvised cat box. At intervals I would go with her to the toilet, put the litter in the box, then put her in the box, but she would just sit in the box and look at me as if I were a crazy person."

The stewardesses were charmed, which is unlikely to happen these days, when they carry plastic handcuffs and rolls of duct tape in their kit to deal with "difficult" passengers. Cleopatra had had her shots, as stipulated by the French Embassy (along with proof of health and proof of ownership, the latter not easy to provide for a rescued stray cat), but the French douanes waved away her documents—*"Une chatte, c'est une chatte, quoi?"*—the French have never been difficult on the subject of bringing animals into France, or out of it, unlike the United Kingdom, which imposes a six-month quarantine on animals. This caused innumerable problems at the very highest level of government when the Duke and Duchess of Windsor proposed to return to England from Madrid with their pugs during the Second World War, and obliged Elizabeth Taylor to live on a yacht, on the Thames, to spare her dogs from being quarantined.

Cleopatra settled down easily into life in Paris, a city whose affection for cats is best represented by the novelist Colette, who adored hers. Referred to by one and all as "Minou!," a kind of generic French affectionate name for cats, much as we tend to call cats whose names we don't know "Kitty," she had the run of Leila's apartment building, from the concierge's lodge to the studio of the Japanese artist who lived downstairs, and would wander out onto the balcony occasionally for a look at Paris street life. Although there *is* such a thing as French cat food, cats in France mostly eat delicious scraps from their owner's plates, the French being at the same time thrifty and unable to believe that any creature would prefer food out of a can to something that has been tastefully cooked in an interesting sauce.

In Cleopatra's case, since Leila was learning cooking, she dined off things like poached wild-caught salmon, and developed as a result a fairly sophisticated Parisian palate for quenelles, kidneys cooked in rich sauces, pâtés, etc.— she developed, in short, a fairly sophisticated taste in food as Leila's cooking skills, already considerable, improved.

There are those who believe that rich food, particularly dishes with a lot of cream or butter, are bad for cats, but needless to say, this view is not held in France, and doesn't seem to be held at all by cats. Most cats *adore* rich food when it's offered to them—after all, *ils ne sont pas si bêtes*—and while it's probably not any better for them than for us, since they don't smoke or drink, and don't live long enough to have to worry about cholesterol, it probably can't do them much harm. Our own cats, though not exposed to French haute cuisine (except when Leila comes for a visit) have, on occasion, eaten cheesecake, key lime pie, chili, a variety of Chinese dishes (when Michael was attending the Chinese cooking course at the Culinary Institute of America), roast pork, moussaka, and shepherd's pie. They don't seem to like things with a lot of tomatoes in them, however, or pasta of any kind. Mumsie was especially fond of butter, which she liked in a good-size blob on a plate, but would eat off a finger, or out of the butter crock if you happened to forget to put the lid back on. She would take margarine, if it was offered, but with slightly less enthusiasm. They definitely prefer cream to milk, and whole milk to 1 percent low-fat milk, but then who doesn't? Mumsie used to rather enjoy a Zabar's raisin-bran muffin as a breakfast treat, in the days when Michael still brought food up from the city, and quite enjoyed a spoonful of scrambled eggs, if anybody was having them. Very fortunately, on the rare occasions when they have been offered a few grains of caviar, after a birthday celebration, they have turned their noses up at it. Too salty? Too fishy? Who knows?

If Leila came to stay more often they would probably

develop more sophisticated palates, but by and large most cats are willing to try good cooking if it's offered to them, which is more than you can say about a lot of people. Of course the amount of faith you can put in the taste buds of an animal that can also enjoy shreds of raw mouse or swallow a large insect whole with every sign of pleasure is questionable.

In any event, Cleopatra thrived on French cooking, and like most Americans who settle down in Paris, soon became *plus française que les français*. During the winter months she developed bright orange patches on her fur, from sitting too close to the gas heater and singeing herself. She came to understand French, since when Leila was away she was sent to stay with a friend who knew no English. The friend would sigh with relief every time Leila returned to pick Cleopatra up, since she lived in fear each day she was away that the cat would die, as she was so old. But Cleopatra never did. She seemed to enjoy being carried through the streets of Paris, absorbing the noise and the strange smells, so very different from those of New York, morphing into *une vraie chatte française,* so much so that Leila worried about how she would adapt when she returned home after five years— a long time in a cat's life, even one so old as Cleopatra.

But her return was uneventful, and she settled easily

into a new routine, commuting between New York City and Connecticut, where the conductors were considerably less cat-friendly than the flight attendants on Air France. Cleopatra would sit on Leila's lap sleeping or looking at the scenery, and seemed to have a sixth sense of when the conductor was coming, at which point she would pop back into her bag and sit quietly, hardly even breathing.

In 1982, she traveled cross-country by air to Los Angeles, where she survived only three weeks before dying. "Who can blame her?" as Leila says—she was thousands of miles farther away from Paris, and Los Angeles just didn't seem to her a civilized city after Paris, a feeling that has been shared by a lot of other visitors, not all of them cats.

Well, it's hard, when you've grown used to looking at the rooftops of Paris through the window, and eating amusing little dishes flavored with garlic and shallots, and hearing French spoken, to end up in a small house in Los Angeles and two meals a day from the shelves of the local supermarket, and poor Cleo, who was by this time twenty-six, a fabulous age for a cat, simply gave up and said adieu to it all.

It used to be said that when good Americans die, they go to Paris, so perhaps it's true for cats as well, and Cleo's spirit roams the boulevards of Paris.

At any rate, one would like to hope so.

10. Cat-harsis: The Cat Life

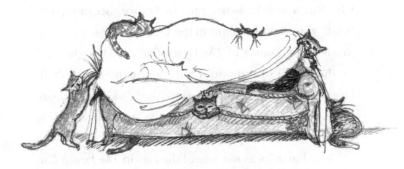

One of the huge, astonishing, and unexpected publishing successes of the late 1950s was a tiny, square, paperback book that sold for a dollar called *The French Cat*, by Siné, a French newspaper cartoonist, who used words including the letters "c-a-t," together with lively drawings, to present both charming art about cats and a challenging word game. Of course, in the original he used the word "*c-h-a-t*" to build on, as in "*entre-chat*" (with a cat drawn as a ballet dancer)—*il était français, quoi?*—but as subsequently adapted into English by Simon and Schuster, it produced words like "cat-astrophe," or

"Cat-alina," or "cat-erpillar," or "cat-sup" (well, you get the idea—if not, see page 141). Everybody in the editorial department of Simon and Schuster sat around thinking of words that contained the letters "c-a-t" for weeks, to the exclusion of much other useful business. As a result, Siné, who was a well-known figure in France, became (very briefly) famous overnight in the United States, and hundreds of thousands of *The French Cat* were bought, and no doubt still sit unnoticed today on many a bookshelf, largely forgotten, or hidden, since the book's modest size tends to make it invisible, rather like a small cat hiding behind the books, appropriately enough.

"Cat-harsis" was not one of the cats in *The French Cat,* since nobody could figure out what drawing of a cat could accompany it, but the word might easily describe a plenitude of cats, or perhaps a quantity of cats beyond which it seems impossible to go. Of course these things are relative. To many people, one or two cats seem like a lot, or too much, to others a half a dozen or so don't seem enough, to a select few, the sky's the limit, the more the merrier.

By and large, it's not the number of *cats* that's the limiting factor—cats don't take up a lot of space, and have a way of vanishing out of sight, except when it's their mealtime—it's the number of litter boxes you're prepared to live with. Our own limit in this crucial area grew from one, to two, to three (plus one outside on the screened-in front porch and another in the barn laundry room, which also don't really count—*inside* the house is all that counts). Three litter boxes seem to represent our own personal comfort limit (until further notice), two downstairs, one upstairs. (Since one of the downstairs ones belongs to Tootsie, who seldom leaves her own private suite in one wing of our house unless she's carried out, limp and howling piteously, and since she also doesn't encourage the other cats to come in, it's really in the nature of a one-cat private bathroom.)

From time to time we have tried more "modern" devices, which appear regularly in upmarket mail-order catalogs as the latest technological advance in cat litter boxes. One of these was what looked like a small fiberglass astrodome or igloo that fit over the litter box, with a hole for an entrance, that was supposed to provide the cat with privacy, and to reduce odors, and eliminate the scattering of litter. We quite liked this, but as luck would have it, the cats didn't like the look of it at all—none of them would go near it, however much they needed to use the litter box. Cats *do* have a need for privacy, but not so much that they wanted to enter a blue plastic igloo and sit in the dark. More elaborate (and expensive) was a high-tech, plug-in, mechanical "self-cleaning" device that could tell when the cat left the litter box, and then, clicking and whirring, combed through the litter with fine metal teeth. This seemed to us like a big technological step ahead of a large slotted spoon, but the cats viewed it with unconcealed distrust, or, in some cases, outright terror.

Well, who can blame them? How many people, after all, really *like* self-flushing toilets (the kind that flush the moment you stand up, while you're still groping for the toilet paper and your trousers are down around your ankles, if you're a man)? I mean, we'd all like to think that we can deal with flushing toilets by ourselves, and cats seem to feel the same way about mechanisms that whir, squeak, comb through the litter, and dispose of whatever is there. They must think: Who knows, maybe I could get scooped up myself? There are exceptions, of course. A good friend of ours, William Steinkraus, the Olympic gold medal–winning rider, has a cat that has learned how to flush the toilet. None of ours has ever grasped this mechanical principle, alas.

Of course we are merely dabblers, amateurs, as it were, in the litter business, much as it sometimes seems to dominate our lives (especially first thing in the morning). Our friends the Lynns—Susan and Jeff—who live on a horse farm up the road from ours, have eighteen litter boxes in their house, down from their maximum number, they explain with a certain degree of sadness, when they had twenty-seven. Twenty-seven litter boxes! We ask the obvious question: Do they empty the supermarket shelves every time they need to resupply themselves with fresh litter? But no, of course not. They shake their heads, like people sharing a trade secret. They order it from Masten's, our local feed store, and have it delivered by truck, on pallets. They have tried everything to make the task easier—clay-based litter, sand-based litter, the kind that clumps and the kind that doesn't, and are now using litter made of tiny balls of old newspapers. Clumping sand was the worst, they agree, since the cats tended to carry it between their toes and leave a thin layer of it all over the pillows and the furniture. Would we like to have a look?

Who could say no? The Lynns have a handsome old house, somewhat more modern and more of a country manor really, with a big formal entrance hall, than our own eighteenth-century farmhouse. On the other hand, their decorating scheme, though they have lived there for several years, seems to have stalled at inception, apparently brought to a halt by the needs of their cats. In the living room, what remained of the carpets has been removed, and the furniture has been covered with sheets—too late, alas, since the cats have already sharpened their claws on every square inch of upholstery. The Lynns have attempted to prevent further destruction by putting up barricades around the more vulnerable corners of the furniture, using what looks like tough, mustard-colored corduroy mounted on L-shaped pieces of plywood, with only mixed results, frankly.

At the entrance to the living room one of the paintings has been removed to make way for a big hook on which to hang a plastic hospital hydration bag and its tubing—several of their kitties have failing kidneys and need to be rehydrated daily. One end of the living room is completely devoted to a long row of litter boxes—the Lynns' cats, unlike ours, have apparently learned to like the kind that have a big plastic cover, like an Eskimo's igloo. In a closet under the elegant curved staircase, is another big litter box, dimly lit with a purple fluorescent light, for those cats who prefer a little more privacy. The small powder room opposite seems to have so many water bowls on the floor that it's hard to see how a person could get to the toilet without putting a foot in water. No cats are in sight, however.

The small, glassed-in porch on which we are sitting has also been stripped bare by, or for, the cats—furniture covered in sheets, a stack of fifty-pound sacks of dried cat food. We chat about the cats' eating habits. Here, the Lynns take a stricter line than we do. Whereas each of our cats gets a different kind of food, and some of them have food cooked for them, the Lynns give every cat the same food, on its own plate—no choices on the menu, no substitutions. Sooner or later, they all learn to like it, or at least to eat it, whether they like it or not. Otherwise, they explain, life gets too complicated.

One can see why. At the moment they "only" have thirteen cats, down from their high of twenty-six. Each of them gets its portrait painted in oils by a pet portraitist, so the Lynns have a painting to remember them by when they're gone.

They didn't start out with the intention of having a lot of cats, Jeff points out—it more or less happened of its own accord. They had one cat, and felt it needed another for company, then, as fate would have it, Sue, a flight attendant, began to fly back and forth to Barbados. Barbados has no rabies, it seems, and like the United Kingdom, of which it is a colony, you can't bring a pet in without a six-month quarantine period. On the other hand, nothing prevents you from bringing a pet *out*.

Barbados, like most of the Caribbean islands, is full of stray cats (having their cats spayed or neutered is not a major preoccupation of the native islanders), so everywhere Sue went she saw cats that attracted her sympathy and attention. First she brought back one, then another—nobody at U.S. customs or immigration seemed to care that she kept on using the same original set of papers for innumerable cats—and before long the Lynns had twenty-six cats and twenty-seven litter boxes in their duplex New York City apartment. Fortunately perhaps, Sue got shifted to another flight, and at about the same time they bought the horse farm, and eventually gave up the apartment, moving their entire cat population up to the country. Since that time, the number of cats has diminished, as the old ones pass on, though the Lynns find it hard not to add the occasional stranger that comes to the door, many of them from a mutual neighbor who, like the Barbadians, also does not believe in spaying or neutering animals.

In various pet-lover magazines there's an ad that shows the potential population growth of a male and a female cat if neither they nor their offspring are neutered. It's rather like those charts that used to show what happened to money if you allowed it to grow at compound interest over twenty years—in any case, before very long, Mother Nature and the reproductive habits of cats produce millions of cats, the message being, of course, to spay and neuter them. Our mutual neighbor seems to be putting this mathematical theory to a practical, experimental test, and is so far producing an almost unlimited number of cats, most of them suffering from some rare and expensive disease, so there's never a shortage for the Lynns to draw on, whenever they feel they don't have enough.

Wally, named because Sue found him as a tiny kitten on a wall near their house and brought him home in her riding helmet, is the kind of cat that would make even a non–cat lover's eyes turn moist. He appears suddenly and silently, apparently from nowhere (How do cats *do* that?), a massive bundle of tawny fur, with huge yellow eyes and a trusting expression. The phrase "long-haired cat" (as opposed to "short-haired") hardly does Wally justice. Even without the fur, he would be a big cat, but with it, he looks immense, so fluffy that he could pass for one of those spun sugar desserts, his silky hair sprouting in every direction. Like certain fat men, he has a kind of elephantine grace, and an almost eerie calm. He is noted, Jeff explains, for his rough tongue. Some cats love licking their owners' faces, Wally being one of them, but his tongue apparently has an abrasive power beyond that of the ordinary cat, which he proceeds to demonstrate as Jeff kneels down to him. Sure enough, Wally licks his cheek, a beatific expression in his eyes, and when Jeff gets up he has a red patch there, as if he had shaved himself too hard with a blunt razor.

"Each of them has a trick," Jeff explains, looking dreamy-eyed himself. "That's Wally's." He thinks for a moment, looking at room after room with the carpets rolled up or removed and the furniture protected as if from imminent civil war (but too late). "The thing is," he says, "they give you something to laugh about every day. Who can knock that?"

Who indeed? Of course, eventually they make you cry too, but that's part of the deal.

As we leave, we see Wally watching me through the glass door of the kitchen—the Lynns don't let their cats out, unlike us. His expression looks as if he might be thinking, "Thanks for coming to visit me," and maybe he is.

After all cats aren't dogs. They don't defend their owners' homes—the idea seems not to have occurred to them. They see the place they live as *their* home, one they share with their "owners," or would if the concept of ownership were clear to cats, but it's not. Just as the Indians couldn't understand how it was possible to buy or sell land—the land was everybody's, it didn't belong to them, they belonged to *it*—cats don't think they're owned by anybody.

Even behind doors and windows, like amiable Wally, they're free. Always.

That may, in fact, be the most important thing about them.